Friendly Interfaces Between Humans and Machines

P. V. S. Rao · Sunil Kumar Kopparapu

Friendly Interfaces Between Humans and Machines

P. V. S. Rao
Department of Computer Science
Tata Institute of Fundamental Research
Mumbai, Maharashtra, India

Sunil Kumar Kopparapu
TCS Research and Innovation
Tata Consultancy Services
Thane, Maharashtra, India

ISBN 978-981-13-4674-3 ISBN 978-981-13-1750-7 (eBook)
https://doi.org/10.1007/978-981-13-1750-7

© Springer Nature Singapore Pte Ltd. 2018
Softcover re-print of the Hardcover 1st edition 2018
This work is subject to copyright. All rights are reserved by the Publisher, whether the whole or part of the material is concerned, specifically the rights of translation, reprinting, reuse of illustrations, recitation, broadcasting, reproduction on microfilms or in any other physical way, and transmission or information storage and retrieval, electronic adaptation, computer software, or by similar or dissimilar methodology now known or hereafter developed.
The use of general descriptive names, registered names, trademarks, service marks, etc. in this publication does not imply, even in the absence of a specific statement, that such names are exempt from the relevant protective laws and regulations and therefore free for general use.
The publisher, the authors, and the editors are safe to assume that the advice and information in this book are believed to be true and accurate at the date of publication. Neither the publisher nor the authors or the editors give a warranty, express or implied, with respect to the material contained herein or for any errors or omissions that may have been made. The publisher remains neutral with regard to jurisdictional claims in published maps and institutional affiliations.

This Springer imprint is published by the registered company Springer Nature Singapore Pte Ltd.
The registered company address is: 152 Beach Road, #21-01/04 Gateway East, Singapore 189721, Singapore

To all those researchers who are striving to narrow the communication gap between humans and machines.

—P. V. S. Rao

To all those who believe in using the current state of technology for larger good rather than procrastinating for the lack of availability of a perfect technology.

—Sunil Kumar Kopparapu

Preface

With the increasing adoption of technology in day-to-day activity, it has become extremely important to build human–machine interfaces (HMIs) that are both natural and convenient to use by all strata of the society. When humans communicate among themselves, whether in the written or oral mode, while the 'messages' are easy to understand for the human observer, the underlying rules of language are not obvious; they are generally complex and tricky to understand or codify. Even in such a tricky scenario, usually humans can quite easily unpack the many nuanced allusions and connotations in every sentence and decode what another person is saying.

Computers are extremely effective in applications involving number crunching; however, things get hard when it comes to using them for language processing applications. In our particular context, being able to make computers 'understand' a question posed in everyday human natural language and respond meaningfully with a precise answer is a sort of the holy grail; this would allow machines to converse very 'naturally' with people, letting them ask questions instead of typing or speaking keywords. In this monograph, we look at the aspects of technologies that are needed for designing systems which allow humans, using natural language, to interact with machines as they would interact with other humans.

The purpose of this monograph is twofold, namely (a) to introduce and explicate the problem of addressing natural language interfaces in a human–machine interaction (HMI) scenario and (b) to discuss how to build usable interfaces through real-life examples. This monograph presents the work done in this area by the authors over a period of time and includes several case studies and examples. Systems developed then are included in the monograph, even though they are of that vintage because of their usefulness in illustrating the principles and techniques which are relevant even now. In the present volume, we do not consider the issue of speech-to-text (automatic speech recognition) or text-to-speech (TTS) conversion. We start with text input from the humans and a text display or output from the machine.

Mumbai, India — P. V. S. Rao
Thane, India — Sunil Kumar Kopparapu

Preface

Acknowledgements

Several people at different points of time have contributed to the contents presented in this monograph, directly or indirectly. It is indeed a pleasure to acknowledge them; however, if we missed out on any name(s), it is purely our inability to recall their names at the time of writing this monograph. Folks whom we would like to acknowledge are as follows (no particular order):

1. Dr. Arun Pande
2. Akhilesh Srivastava
3. Dr. Nirmal Jain
4. Charles Antony
5. Rekha Joshi
6. Swapna Sundaram
7. Rimzim Sinha
8. Umesh Kadam
9. Meenal Orkey
10. Dipti Desai
11. Meghna Pandharipande
12. Prakash Passahna
13. Irfan Khan
14. Anish Mathew
15. Ambhikesh Shukla
16. Dr. Gian Sunder Singh
17. Sumitra Das
18. Byju Mathew
19. Ambhikesh Shukla
20. Joshua Pinto
21. Mustafa
22. Kalyan Godavarthy
23. Aditya Mishra

Contents

About the Authors

Dr. P. V. S. Rao is an Indian computer scientist, known for his research in the fields of speech and script recognition, who has contributed to the development of TIFRAC, the first indigenously developed electronic computer in India. He is a recipient of various awards, such as the IEEE Third Millennium Medal, the Vikram Sarabhai Award, the Om Prakash Bhasin Award and the VASVIK Industrial Research Award. The Government of India awarded him the fourth highest civilian honour of Padma Shri in 1987. He graduated in science from Utkal University in 1953 and received a master's degree in science (physics) from Banaras Hindu University in 1955. He served Tata Institute of Fundamental Research (TIFR) in the Department of Computer Science from 1955 until retirement. He subsequently built the Cognitive Systems Research Laboratory in the capacity of a research advisor with Tata Infotech. He was also a research advisor to Satyam and Tata Consultancy Services. He obtained a doctoral degree in physics from the University of Mumbai for his work on the display of text and graphics on computer screens.

Dr. Sunil Kumar Kopparapu obtained his doctoral degree in electrical engineering from the Indian Institute of Technology Bombay, India, in 1997. His thesis 'Modular integration for low-level and high-level vision problems in a multi-resolution framework' provided a broad framework to enable reliable and fast vision processing. In his current role as Principal Scientist with the TCS Research and Innovation - Mumbai, he is actively working in the areas of speech, script, image and natural language processing with a focus on building usable systems for mass use in Indian conditions. He has also co-authored a book titled *Bayesian Approach to Image Interpretation* and more recently a Springer Brief on Non-linguistic Analysis of Call Center Conversation and Analyzing Emotions in Spontaneous Speech in addition to several journals and conference publications. He also holds a number of patents.

About the Authors

Abbreviations

AI	Artificial Intelligence
ASR	Automatic Speech Recognition
FAQ	Frequently Asked Question
HHI	Human–Human Interaction
HMI	Human–Machine Interaction or Human–Machine Interface
KC	Keyconcept
KC-KW	Keyconcept-Keyword
KW	Keyword
ML	Machine Learning
NLP	Natural Language Processing
QA	Question Answering
QAS	Question-Answering System
QU	Question Understanding
SE	Search Engineering
SMS	Short Messaging System
SS	Syntacto-Semantic
TTS	Text-to-Speech
UI	User Interface
UNL	Universal Networking Language
UWs	Universal Words
Ux	User Experience
WSD	Word Sense Disambiguation

Abstract

The recent decades have seen phenomenal advances in information technology (IT) and its applications in diverse fields, including those which seemingly require human intelligence. This underlines the need for powerful human–machine interaction (HMI) systems. Children learn (human–human) interaction skills spontaneously and gradually by observation and by actual interaction with parents, family and other children as well as in larger social settings. Given that HMI is the need of the hour, a need spurred by technological changes in the society, the importance of developing powerful and easy-to-use HMI systems cannot be overemphasized.

In this monograph, we specifically focus on human–machine interactions with particular emphasis on making them user-friendly and, as an ideal goal, as natural as human–human interaction (HHI).

For realizing this, firstly, we need to understand the expected behaviour of the person interacting with the machine. Secondly, we need to recognize and appreciate technological limitations that constrain implementation of systems that make HMI both possible and usable.

At present, on the one hand, researchers keep focusing on bettering the technology without necessarily worrying adequately about its usability; on the other hand, implementers and developers remain primarily interested in building quick solutions, without worrying about the capabilities and limitations of the current state-of-the-art technology.

In this monograph, we try to take a middle path:

1. We build up a clear understanding of various aspects of natural language.
2. We keep in sight the power as well as the limitations of the available technology.

We hope that our approach will enable developers to build human–machine interfaces that are easy to implement and use. Hopefully, it will also help the beginner student to obtain a clear understanding of the subject in a proper perspective.

To start with, we review the evolution of language as a spontaneous natural phenomenon in the overall scheme of the evolutionary development of living beings. Such a study is necessary and will be very useful in the context of attempting to use natural language as a convenient medium of communication between humans on the one hand and machines on the other.

We then examine possible approaches to understand and represent the meaning and the common aspects of human–human and human–machine interactions. We introduce our keyconcept-keyword (also called minimal parsing) approach as a convenient and realistic way to implement usable HMI systems. We also describe a *working* question-answering (QA) system along with a number of practical applications based on our minimal parsing approach.

Chapter 1
Dissecting the Monograph

Interfaces are the touch points between two or more entities which do not 'speak the same language' or work in ways that are dissimilar. A classical and extremely relevant example is the interaction between humans and machines. Figure 1.1 captures the difference between humans and machines figuratively. The human is shown as a circle (to represent versatility and variability) and the computer appears as a rectangle (to represent task-specific functionality). Thus, there is a gap between the two. Interfacing technology provides mechanisms that allow this gap to be bridged, thereby making it possible for these two different entities to communicate or talk to each other.

There could be several ways of achieving this connection between the two entities (e.g. the longish red line — path or the shortest blue line —..—..). The length of the line connecting the two entities is an indicator of how 'well' the interface connects the two entities. For example, we could have an interface that is very machine friendly and allows the human to interact only in a constrained manner (the long red line —); on the other hand, there could be an extremely human-friendly interface which allows the human to interact with the machine in a manner that is natural (and easy) for him, without being limited by any constraints arising from limitations of the machines (the blue line —..—..).

In principle, both these interfaces would allow one to 'connect' the human and the machine; however, neither would be optimal or good. A *usable* interface would be one that allows the human to interact as naturally and freely as is feasible within the realm of current technology. In practical terms, it is presently not realistic to aim for completely unconstrained human interaction. The implementer's aim should therefore be to realize a user interface that is *usable*; i.e. minimal constraint on the manner of human interaction, within the limitations of the contemporary state of the art in technology. A user-friendly interface would give more importance to the needs of the human in the loop (making it easy to use) by stretching the interface technology to its limit and making the whole idea of the interface between a human not just usable but also convenient.

© Springer Nature Singapore Pte Ltd. 2018

P. V. S. Rao and S. K. Kopparapu, *Friendly Interfaces Between Humans and Machines*, https://doi.org/10.1007/978-981-13-1750-7_1

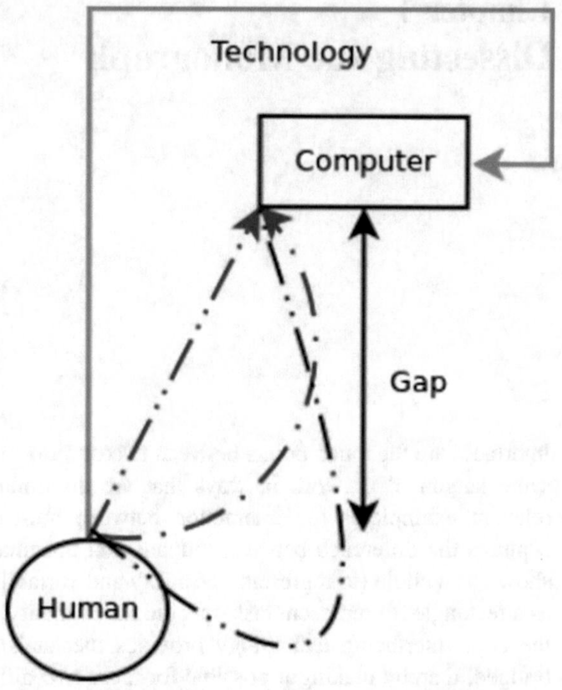

Fig. 1.1 Human–machine interaction possibilities

For example, consider a voice-based self-help system. A voice-based interface (shown in Fig. 1.2) where humans can interact with the machine and help themselves. Figure 1.2 captures a typical transaction between a human and a machine; the user speaks into his mobile phone and asks /`When was my last commission paid`/. This audio query goes through the mobile service provider gateway and lands on a speech recognition server which converts it to the text string '`when was my last commission paid`'. This text string is passed onto the natural language server to understand the intent and then to process it to get the desired information, in this case, the date 12 May 2007. Now, this answer is sent as a short message (SMS) back to the user on his mobile phone. Alternately, it can be read back to the user on his mobile phone as a spoken message.

The most desirable option would be to allow the human to speak freely to the machine (i.e. allow the speech recognition system to have a free grammar) for doing his day-to-day transactions; but then this would possibly cause higher speech recognition errors (the blue short line; Fig. 1.1). On the other hand, one could permit the human to only query the system in a certain prescribed manner (i.e. constrained grammar), this is the red line path (see Fig. 1.1).

However, neither of these provides a good user experience (Ux). The middle path would be to allow the speaker to query without constraints but to have a system in place that is able to decipher the *intent of the query* (usable grammar) rather than having to recognize and fully understand the actual spoken query itself.

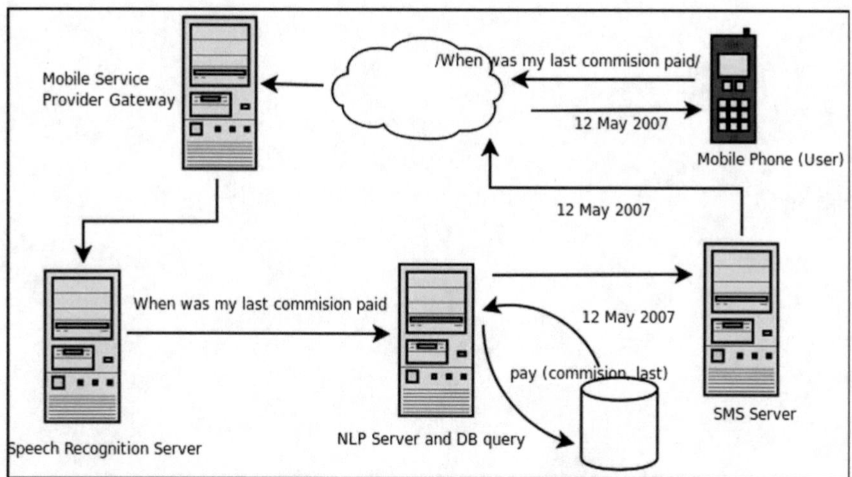

Fig. 1.2 Block diagram of a self-help system

This example illustrates the concept of usable technology. This is however true in all situations where there is a need for an interface to 'connect' human and machine. This monograph deals with one such interface: it allows the human to interact with the machine in a way which is natural but allows his interaction to be meaningful and friendly, by adopting appropriate technology for the interface.

Chapter 2
Introduction

User interfaces enable humans to communicate with machines. We may call such systems human–machine interaction (HMI) systems. Human communication with machines could be in one or more of several possible modes: text, speech, gesture and so on. In this monograph, we are primarily concerned with communication using text.

HMI systems would have three main components: the human, the computer and the information (knowledge) bank which the computer would need to use for usefully achieving the objective of the human who is interacting with the machine. The interface between these (part hardware and part software) ensures the effective operation of the entire system.

In this chapter, we touch upon aspects of language that are relevant to the discussion in the rest of the monograph. We first give a very brief description of natural language and then talk of formal language. We briefly describe the process of looking at natural language using the formal languages to give a feel for natural language processing.

2.1 Language

In cognitive science, the word 'language' refers to the human faculty or facility of creating and using language [1]. Languages use finite numbers of symbols (ideograms, letters, phonemes, words or gestures) to represent (or encode) and understand (decode) information being conveyed. The encoding and decoding have to be in a (more or less) consistent manner for each language for reliable transfer of information from the sender to the receiver.

By virtue of evolutionary development, most languages are based on audio communication; the sounds or sequences of sounds used, however, do not necessarily have intrinsic or obvious meaning. They have, over a period of time, come to be associated with generally *agreed* meanings. Some linguists claim that this began

© Springer Nature Singapore Pte Ltd. 2018
P. V. S. Rao and S. K. Kopparapu, *Friendly Interfaces Between Humans and Machines*, https://doi.org/10.1007/978-981-13-1750-7_2

with the gestures and sounds involved in expressing emotions, feelings or activities and these came to be associated with and to represent the intended meaning. The gesture of pursing the lips involved in infant suckling milk from its mother's breast and the associated sound of */m/* came to represent the concept of motherhood—and the words *m*aa, *m*ater, or *m*other—in most languages. Language developed and evolved as new words were invented and added to the vocabulary and rules for stringing the words together developed over long periods of time in diverse locations and under varying conditions. Languages that evolved spontaneously and naturally in that manner are called *natural* languages, in contrast with, for instance, programming languages that have been designed by specialists for specific purposes over short periods of time. As written scripts evolved much later than speech, spoken languages predate written ones. In fact, several languages do not have written versions even today.

It should be noted that machines use artificial entities called formal languages (such as programming languages) which are used to cause them to perform specific tasks. Like natural languages, programming languages are also encapsulated by syntactic and semantic rules which give an agreed upon structure and meaning respectively to the programming language. In a very coarse way, there is a need to try and understand the complexity of the natural language, either textual or spoken, using formal languages to enable humans to interact with machines naturally and in a meaningful way.

2.2 Natural Language Interfaces

Language usage, in general, can be very tricky especially in a country like India because of the large socio-economic gap in the population, diverse cultures, large number of languages and finally, the difficulty in entering Indian languages electronically using keyboards or touch screen. The complexity in building interfaces is compounded for spoken languages because of the additional complexity of the existence of a large number of dialects, language accents, use of multiple languages in one sentence and the noisy communication (telephone) channel. However, humans are able to communicate very effectively with each other even in cases where they are illiterate or minimally literate. Conversations often contain incomplete or even ungrammatical sentences, but that does not seem to significantly compromise the effectiveness of communication. This seems to indicate that complex grammars or detailed parsing are not absolutely necessary to understand the meaning of sentences in human-to-human interaction and that there is at play a simpler mechanism than actual grammatical parsing. We propose such a mechanism which we call as keyconcept-keyword (also called minimal parsing) approach.

Considering that natural language interaction between humans and machines ought to ideally be as close as possible to typical human-to-human interaction, we might expect that our above-mentioned understanding of human–human communication will be useful in designing natural language interfaces and as we will show in this monograph that it turns out to be so. These aspects are dealt with in adequate detail in later chapters of the monograph.

Also, designing interfaces that can effectively communicate with people require flexibility in terms of being able to understand the *intent* and hence the central theme of the query. Understanding language semantics enables the interface to fetch multiple entities in a single natural language query rather than making the user seek these information entities in stages in order to get sufficient information in answer to the query. This not only makes the interface natural to use but also convenient.

For example, in a typical speech-based query system currently being used in banking, the system makes the user walk through a predetermined menu or tree, in the manner of earlier keystroke-based telephone interaction systems. First, it asks for information to ascertain if the query is 'savings account' or 'current account' related; then, depending on what the user said, it would seek the next detail of information entity—say, 'account balance', 'cheque book request' or 'stop cheque request', etc. This can be quite annoying and is definitely not the natural way in which a human would ask for this information. In contrast, in case of a natural language (and speech-based) system, the user would have the flexibility to ask—/`What is the balance in my savings account`/ or /`Could you please let me know my savings balance`/, etc. The ability to answer such varieties of queries without ambiguity is possible when (a) the domain of transaction is well defined (banking, travel, etc.) and (b) the HMI system is able to analyse a natural language query and respond in a way another human would have. As we mention elsewhere in the monograph, the system should be able to respond with *feel* answers rather than *reason* answers.

Question answering (QA) systems are an instance of HMI. In general, QA systems are able to understand a query posed in natural language and respond with a precise, factual answer [2]. In other words, QA systems must do more than what search engines like Google [3] and Bing [4] do, which is merely point to a list of hyperlinked documents where you 'might' find the answer. Unlike generic search engines, QA systems have to identify the correct answer on its own. Technologists have long regarded this sort of artificial intelligence (AI) as a holy grail, because it would allow machines to converse more naturally with people, letting people ask questions instead of merely typing or speaking keywords or keyphrases. The reader should note that this is very different from Deep Blue [5], for instance, was able to play chess well because the game is perfectly logical, with fairly simple rules; it can be reduced easily to mathematical and logical computing.

However, the rules of natural language are much trickier. As humans, we can easily decode what someone else is saying, and we can effortlessly and easily unpack the many nuanced allusions and connotations in every sentence. To understand natural language the way humans understand and interpret is hard for computers because natural language is full of intended meaning. The very best QA systems, even today, could sieve through news articles on their own and answer questions about the content, but they understand only factoid-based questions [6] stated in very simple language (for example '`What is the currency of India?`', '`How many medals did India win in Asian games`', etc.).

References

1. Wikipedia, Language [Online]. Available: http://en.wikipedia.org/wiki/Language
2. D. Jurafsky, J.H. Martin, Question answering, in *Speech and Language Processing*, https://web.stanford.edu/~jurafsky/slp3/ed3book.pdf. Accessed Feb 2017
3. Google, Google search [Online]. Available: http://www.google.com
4. Microsoft, Bing search [Online]. Available: http://www.bing.com
5. https://en.wikipedia.org/wiki/Deep_Blue_(chess_computer)
6. P. Ranjan, R.C. Balabantaray, Question answering system for factoid based question, in *2016 2nd International Conference on Contemporary Computing and Informatics (IC3I)*, Noida, 2016, pp. 221–224

Chapter 3
Language

It is remarkable that grammars of all languages have common features at a deep level as well as at the structural level. This might mean that language originated at one geographical location and then spread all over, changing radically as it spread. On the other hand, it can also be argued that this commonality (or similarity) at the deep and structural levels of languages is because language development is dictated by brain structure and this brain structure is a common factor across different groups of humans who might have developed linguistic competence at widely distributed geographical locations. The opinion on which of these two is true is divided.

3.1 Evolution of Language

Language is a primary means of conveying information between humans [1]. Language capability was acquired by humans as an evolutionary development. It would be instructive and interesting to study the evolution of language as a natural phenomenon in the overall scheme of evolutionary development of the world as we know it now. Such a study would be very useful in the context of attempting to analyse the use of natural language as a medium of communication between humans on the one hand and machines on the other.

It is well known that evolution is inefficient, slow and random. It is not 'driven' by any external agent or with an objective towards a predefined goal but inches towards perfection in a relentless process of optimization by trial and error.

Talking about language in particular, interpreting the word in a wider sense, it has been the main means of communication, it exists even in organisms like ants [2] and bees [3]. It is significantly more complex in dolphins [4], whales [5] and higher primates. Language as a mode of communication is comparatively restricted in non-humans as well as in early humans. Communication was initially a combination of sound and gesture and this required *line of sight* communication. Eventually, however, this progressed to a level where gestures became secondary and sound

© Springer Nature Singapore Pte Ltd. 2018

P. V. S. Rao and S. K. Kopparapu, *Friendly Interfaces Between Humans and Machines*, https://doi.org/10.1007/978-981-13-1750-7_3

(speech) became the primary carrier of information. This removed the *line of sight* limitations for intercommunication in humans.

As mentioned earlier, there is no unanimity about whether, during the course of human evolution, (a) 'language' evolved at one single location and spread worldwide along with migration of humans or (b) it developed independently at different geographic locations among different groups of humans. It is clear, however, that the ability to use language marked a major landmark in human development. Groups that had some sort of (speech and hearing based) language could communicate among themselves over distances without any *line of sight* constraints. This meant that these groups could orchestrate and organize hunts and fights much better than others who did not have this ability. Therefore, they could dominate over and eliminate quickly (on an evolutionary timescale) the less fortunate groups that did not have language capability.

Since language provides the means for better organized thinking, linguistic ability facilitated intellectual development. Intellectual development in turn led to the evolution of richer and more complex, hence more effective languages, leading in turn to better intellectual development. The development of speech, language and intelligence by humans happened in a virtuous cycle of mutual reinforcement, and hence, the rapid development of intelligence. Once these three reached a minimally adequate level of self-sustenance, like a nuclear reactor going critical, 'human' intelligence took control of language development, and this progressed much faster than the usually seen evolutionary pace. This virtuous cycle of development of speech, language and intelligence is a remarkable stage in human evolution.

It can be noted that languages and grammars have evolved together, progressively gaining complexity till presumably they reached an optimum level for effective human communication. An interesting point to re-emphasize in this context is that evolution of language in a spontaneous manner stopped when it was "barely" adequate to "deal with real world" and could sustain the intellectual development of the individual and that of groups of individuals. *Higher levels of complexity merely amount to verbal gymnastics.* At this stage, humans seem to have taken control of the development of language by taking an active part in changing (improving) it. As a consequence, language is not optimized to the same evolutionary degree of perfection as several other faculties—e.g. sight and smell—of various species.

The linguistic diversity amongst different languages in the world happened due to wide migration and mutual isolation. As the species adapted to different environments, quick fixes, cultural flavours and creative idiomatic usages all lead to diversity and divergence. Etymological similarities between individual languages arise due to common origin from individual mother languages. As we mentioned already, the fundamental similarity between the different languages and language groups (e.g. parts of speech, broad sentence construction) may even be due to the fact that we all have the same genetic heritage or baggage—the internal hardware in the brain that deals with language.

3.2 Grammar

Language can describe complex (and dynamic) relationships between entities. Language can be used to ‘draw pictures’ using words (but not always too well; hence the adage, ‘a picture is worth a thousand words’). Language is commonly expressed through speech. Due to limitations of human processes and of the media of representation and communication, both speech and written language are constrained to be linear strings of words. Grammar helps us to get over the limitation of linear representation of complex ideas.

Grammar makes it possible to string words in a particular manner to represent the ‘picture’. It is also needed to interpret the string, extract the structure in the construct and recreate the intended ‘picture’ (in the mind of the listener). Take, for instance, the sentence:

“The boy who wears a cotton shirt ate a green apple in the kitchen”

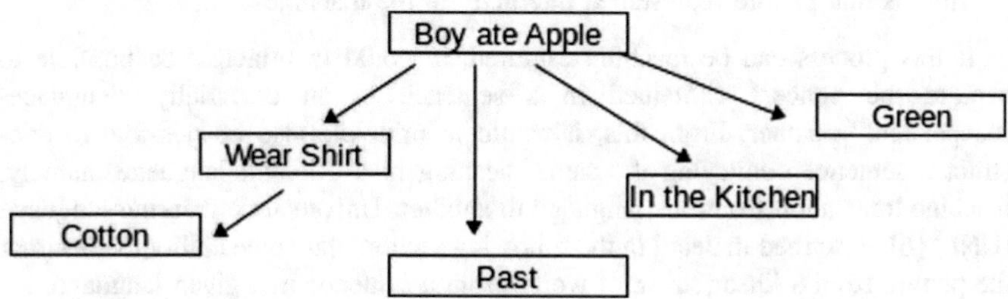

The intention is to convey semantic relationships between the objects or entities that lexical words represent. The relationship is represented or elicited by means of the syntactic relationship between the words, as dictated by the rules of grammar which establish syntactic links between words. In effect, syntactic links are used to indicate semantic relationships.

Thus, grammar is needed also to interpret the string and recreate (in the mind of the listener) the picture that is intended to be conveyed. For example,

`Boy ate apple`

- Syntax
  - noun, verb, noun
  - subject, verb, object
- Semantic implication
  - perpetrator, act, target

```
Cotton shirt
```

- Syntax
  - adjective, noun
- Semantic implication
  - quality, object

The human uses syntax and semantics to extract the richly structured meaning from a linear string of words and is therefore able to visualize in their mind a picture of “the boy”, “the shirt” he wears and the act of “his eating” a “green apple”.

We now have established the background that is necessary to appreciate the problem that a researcher faces when he attempts to use natural language for communicating with a machine. The questions to ask are:

1. Given such a natural language (linear string of words) sentence, how does a machine construct such a mental picture and
2. How is that picture represented internally in the machine?

If this process can be machine captured, it would in principle be possible to capture the concept contained in a sentence in an essentially “language-independent” manner. From this, it would in principle also be possible to construct a sentence conveying the same meaning in a different language, namely, machine translation from one language to another. Universal networking language (UNL) [6], described in detail in the following section, has some ability to construct the picture from a linear string of words from a sentence in a given language.

3.3 Richness of Language

Any concept that one can conceive of can in principle be conveyed by the use of language. Even complex concepts can be communicated using simple means. Observe that linguistic communication is, as we saw, restricted to linear strings of a small number of symbols—a very small number of characters (text; e.g. 26 alphabets in English) or phonemes (speech; e.g. 44 phonemes in English)—using a finite set of words (vocabulary). However, this finite and small symbol space imposes no limit to the variety or the complexity of the concept to be communicated. Consider the complex concept being conveyed by the liner string of words in the following sentence:

```
The tall boy who ate a big mango yesterday, ran quickly to the
station.
```

The power that grammar provides can be seen in the complex construct of a sentence (it might require several scans, even by a human, before it is understood) as shown below:

`The man who, at one time, was a common, undistinguished person with legal training which was rather limited, had the misfortune of being thrown out of a train by an egoistic fellow traveller who was scandalised that a person with the wrong colour of skin is in the same compartment as himself, but the consequences of this seemingly insignificant incident had ramifications which changed the history of two countries and even the manner in which humanity looks upon age-old concepts such as resistance to authoritarianism, picking up seemingly insignificant issues to make a very significant point, integrating a large geographical region into a united mass of humanity with unprecedented unity of purpose, and eventually accomplishing a major political and sociological reform, without a drop of blood being shed.`

There can, of course, be tricky situations: for instance, consider the sentence:

`The newspaper said the police is responsible for violence in the neighbourhood.`

This gives two totally different meanings depending on punctuation (or pauses, in case of speech), namely,

1. `The newspaper said, the police is responsible for violence in the neighbourhood.`
2. `The newspaper, said the police, is responsible for violence in the neighbourhood.`

3.4 Universal Networking Language [6]

Universal Network Language (UNL) is a formal language that has been developed at the United Nations University. It is intended to represent the meaning of natural language sentences in a language-independent form, while preserving all the information contained therein (this means the reverse transformation would also be possible without information loss). This representation is language independent: i.e. independent of the language of the source sentence; it is also machine independent. This means that information can be extracted, stored, retrieved and propagated uniformly for all languages across all machines as long as an encoder (language to UNL) and a decoder (UNL to language) exists for that language. It is easy to see that such a language (UNL) would be useful as an *interlingua* for machine translation of languages; if sentences in the source language are first converted into UNL, these can in turn be converted into the destination language. UNL can also be useful for language-independent representation and handling of knowledge sourced from different languages. Retrieval in any language would be possible and fairly straight forward, irrespective of the source language. Even in case it turns out to be not possible to preserve every nuance, it would be quite adequate for capturing the essence or core of the intended sense, as required for normal person-to-person communication in everyday situations.

Sentences in UNL consist of universal words (UW) linked by relations between pairs of UWs. UNL characterizes each word using several different attributes: nature, genre, syntactic and semantic links and its grammar. UWs characterize universal concepts, written in natural language, accompanied by additional information as attributes which help for example to disambiguate in case of words with multiple meanings (for example, the word "pen" could be a writing instrument or an enclosure for keeping pigs). Separately, there is a database of UWs that are organized so as to link related concepts appropriately (e.g. cow and animal—instance of; leg and cow—part of; flower and blossom—equal to). In this context, the reader may also see the section on WordNet later in this chapter. UNL representation can incorporate the speaker or writer's emotion or intent (surprise, disbelief, interrogation, respect and so on) also.

Conversion from a natural language sentence into a UNL representation involves including attributes and relationships and requires human involvement in an interactive fashion; hence this is a semiautomatic process. Conversion from UNL to a destination language can be fully automatic.

The objective of UNL was that it should be useful in e-commerce, medicine, social welfare, business, libraries and entertainment as well as speech recognition and synthesis. It was hoped that UNL can connect—and even improve—all kinds of human activities.

3.5 Transactional Language

Our discussion so far has been concerned with the grammatical language that one encounters in printed text or formal speech. Quite often, however, the kind of sentences and phrases that are used for purposes of transactions between humans is neither very complex nor very simple. In many cases, they are grammatically challenged; sometimes the remarks are incomplete. For example, the string

```
Two tickets to Manali
```

is ungrammatical; it is also incomplete in terms of when the ticket has to be booked, by which mode of transport or by which class! Yet, such remarks are accepted and with some clarifications sought and provided, effective in accomplishing the task at hand. One would prefer this flexibility to be available in HMI systems. It would therefore seem that grammar in the traditional sense may not be the most appropriate tool for use in this context. We shall return to this topic in later sections.

There is another question that arises in this context: is syntax alone likely to be adequate for HMI systems? Is it not necessary to incorporate some level of capability for understanding the meaning of the words, phrases and sentences that one encounters here?

A simplistic approach to implementing a HMI system would be to visualize it like a straight forward search engine: take the individual words in the query by a user and search the knowledge-base for occurrence of this group of words in a sentence or a passage. One may or may not take into account the sequence in which the search words occur in the query. The limitations of such an approach are obvious.

An improvement would be to take into account not merely the set or sequence of words but try to consider the *concept* underlying the query. Grammar, as we briefly saw earlier, helps in this context. Even that, however, does not take us far enough, because we are concerned with meanings of the word used, and not the word per se. For instance, the query "`who killed Kennedy`" would fail to get a response if the knowledge-base contained the sentence "`Oswald assassinated Kennedy`". The fact that `kill` and `assassinate` signify the same act in this context would be lost, unless that is specifically provided for as information.

There are other similar but even more complex relationships that exist between words, and it would be desirable to incorporate this knowledge into the HMI system, e.g. the relationship between `long` and `short`, `animal` and `cow`, `cow` and `leg`, `eat` and `drink`, to mention a few. The human child acquires this capability to see such relationships, partly by actual teaching but largely by multisensory observation and contemplation. The ability to realize these relationships seems intuitive and natural; the human tendency is to impose order on facts and observations; something like group individual instances into classes and group these classes together to fit this concepts into a hierarchic organization of increasing generality towards the apex.

It is this ability that enables a shopkeeper, if he is asked for "`a tube of Pepsodent toothpaste`", to respond for instance that he can instead supply a "`tin of Colgate toothpowder`". It would obviously be very desirable to provide such an ability in an HMI system.

3.6 WordNet

WordNet can be thought of as a repository, usually handcrafted, that captures the semantic relationship between words. In the previous section, we saw the importance of incorporating such *semantic* relationships between words into HMI systems. WordNet, created in the Cognitive Science Laboratory of Princeton University [7], is a representational system which incorporates and integrates such relationships, in the form of very comprehensive sets of networks or trees of words. These trees (which we call *taxonomy trees*) represent relationships between words on the basis of their sense or meaning. If a word has two or more meanings (e.g. "bank" as in river bank or money bank) it would occur in the appropriate number of taxonomies.

WordNet is in fact designed on the basis of our concepts of how the human memory itself operates. What follows is a brief and partial description of WordNet, to provide the reader a feel for the system. WordNet is a database that incorporates several features which are very relevant for us. Some of these would be useful to clarify our understanding of the type of relationships that might exist between words while some are in fact most appropriate for actual incorporation into HMI systems. This will become quite obvious as we go along with our study of WordNet.

WordNet incorporates the symbolic representation (lexical form) as well as the associated concept, sense or word meaning. It is quite comprehensive and includes most words in the English language. Note that language-specific WordNets have been built, for example, Hindi WordNet [8]. The lexical words in a WordNet are grouped into synonym sets. Each set represents one underlying lexical concept or sense of the word. The linking between these sets happen in the form of various relationships. WordNet is thus essentially a highly connected graph where the nodes (words) are connected by the relationships between them which are the edges of the graph.

Words are organized in synonym groups of nouns, verbs, adjectives, and adverbs. Function words such as "of", "under" and so forth are not included. Let us now look at some of the basic features of WordNet.

Synonyms are the most important underlying basis for WordNet. Two words are *antonyms* of each other if they are opposites (e.g. "heavy" and "light"). A word is called *polysemous if* it has two or more meanings (e.g. "bank" which we mentioned earlier). A word "Wa" is a *hyponym* of another word "Wb" if "Wa" is *a kind of* "Wb" and "Wb" is a *hypernym* of "Wa" ("cow" is a hyponym of "animal" and "human" is a hypernym of "woman"). In our earlier example, `toothpaste` and `toothpowder` are hyponyms of tooth-cleaner, A word can have several hyponyms. This concept is therefore a very useful one for HMI systems.

A word "Wa" is a *meronym* of word "Wb" if it is *a part of* "Wb". The word "leg" is a meronym of "cow" because it is part of it. "Wb" is the *holonym* of "Wa". This relationship is asymmetric. It is also non-unique because a meronym may have several holonyms.

WordNet has 25 *categories* for nouns, such as {act, action, activity}, {natural object}, {animal, fauna}, {person, human being}, {relation}, {feeling, emotion}, {state, condition}, {group, collection}, {substance}, {location, place}, {time} and {motive}.

Adjectives are organized into classes such as *descriptive* ("large", "nice"), *relational* ("joint", "unitary"), *reference modifying* ("earlier", "suspected") and *colour* ("white", "green", "dark"). Descriptive adjectives admit of the following relationships: *antonymy, gradation, markedness, polysemy and selectional references.*

Gradation refers to a *gradual change in meaning* from one extreme to another; some members, which are far apart from each other, may even be antonyms of each other, e.g. "astronomical", "huge", "large", "standard", "small", "tiny", "infinitesimal"; and "superb", "great", "good", "mediocre", "bad", "awful", "atrocious".

Verbs are *most important* because a sentence cannot be grammatical if it has no verb.

The semantic relations used for nouns and adjectives are applicable for verbs also but with some modifications. Polysemy (having more meanings) is much greater for verbs, as compared to nouns. (For example, I beat the system, him, or an egg, etc.). Exact synonyms are rather few, similarities being largely a matter of degree.

There are two ways of organizing verbs: top-down (dividing them into categories and subcategories) or bottom-up (starting with individual verbs and grouping them depending on mutual relationships). WordNet groups verbs into 15 categories: Bodily care and functions, Change, Cognition, Communication, Competition, Consumption, Contact, Creation, Emotion, Motion, Perception, Possession, Social interaction and Weather.

An important relationship used for organizing verbs is entailment (e.g. "breathe" entails "live"—a dog is breathing means it is living). Recall that entailment in verbs is similar to meronymy in nouns, that we discussed earlier. Temporal inclusion is when a sub-activity is part of an activity (e.g. "ploughing" is a sub-activity of "farming"). Decomposition consists in dividing an activity into a sequence of sub-activities (painting a wall involves scraping off existing paint, sandpapering, applying putty, primer, coats of paint, etc.).

Troponymy is a special type of entailment (e.g. stutter and talk—stuttering is a particular type of talking, it entails and is coextensive with talking.)

Verb taxonomies: Organization of verbs is much more complex than that of nouns. There are several individual hierarchies for each category. The verb "move" has two, namely, (a) move a body part and (b) move oneself from location "La" to location "Lb". Verbs of possession have three, namely, give, receive and hold. Communication verbs are of two kinds: verbal and non-verbal.

Overall structure: It would be useful to examine the manner in which all this information is organized in WordNet . The following graph indicates the individual synsets and the pointers that connect them, i.e. the manner in which they are connected. These pointers can be reflexive. Hypernym is a reflect of hyponym because of "Wa" is a hyponym of "Wb" then "Wb" is the hypernym of "Wa". Holonym and meronym are similarly reflexive, while antonym is the reflection of itself.

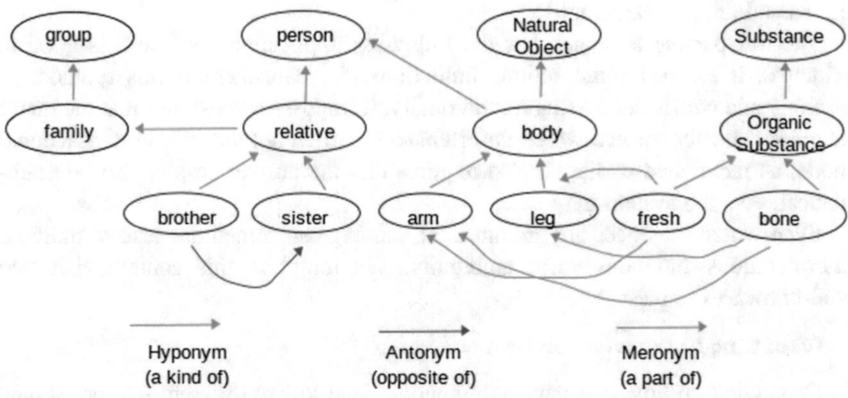

The above discussion helps to get an idea about the kind of relationships between words that it would be desirable to incorporate into an HMI system to give it flexibility in cases where an exact answer is not available. This as we will see is very important to give an answer with a “feel”.

3.7 Search Engine | Query Answering Systems

Typically, in most query answering systems, the input query is preprocessed, appropriate parameters are extracted and these are matched against the corresponding parameters in the answer paragraphs. To permit this comparison, similar parameters need to be pre-extracted from answers (in the knowledge-base) also in a similar fashion.

There are several ways of doing this. At one extreme, one could pick selected words called keywords/keyphrases from the query and match these with similarly selected keywords (KW) and keyphrases extracted ab initio from answer paragraphs. This approach however has serious limitations. Such a system which accepts queries in natural English does not understand the intent, being just based on the lexical words in the natural language query because it is a KW-based search. To a question

“`How does a laptop work?`”

the system might display answers which might talk about “How stuff works”, or “How the mind works”. Note that in this scenario we are speaking about is a pure keyword-based search, where we look at each lexical word as a standalone word and pick an answer which contains most of the keywords in the question.

And at the other extreme, one could fully parse the query, identify the parts of speech (PoS) of each word, and then match the parsed structure with structure of the pre-parsed answer paragraphs.

The full parsing approach has the following implications: While it is good in principle, it too has some serious limitations: for one, there is this problem of accurate and consistent parsing. Additionally, it imposes a constraint that the query be grammatically correct, which unfortunately is often not the case in transactional mode, as mentioned earlier. It also requires that the answer corpus also be grammatical; else the system may fail.

Even where sentences are grammatical, parsing can sometimes lead to multiple interpretations and consequent ambiguity. We might in this context visit two well-known examples:

“`Visiting Relatives can be a nuisance`”

Depending on how it is parsed, this would lead to two different interpretations, namely,

- Relatives (who visit you) can be a nuisance
- (your) Visiting (your relatives) can be a nuisance

The sentence

“`Time flies like an arrow`”,

has several possible interpretations.

- Time flies [9] (a special kind of flies) have a fondness for an arrow?
- Time (as in space-time) flies (fast) like an arrow?
- Time flies like an arrow? (time the flies—with a stopwatch—as you might time a race horse)

It can be argued that full parsing is not required in general for a question answering system. One can for instance use good heuristics to get a correct or an approximately correct answer. As we will elaborate in the following chapter, a minimal parsing (or keyconcept-keyword) approach is eminently suitable for this. In this approach, there is no explicit constraint regarding grammar.

References

1. http://ro.ecu.edu.au/cgi/viewcontent.cgi?article=7526&context=ecuworks
2. D.E. Jackson, F.L.W. Ratnieks, Communication in ants. Curr. Biol. **16**(15), R570–R574
3. P. Borst, http://www.beeculture.com/communication-among-bees/
4. M.O. Lammers, J.N. Oswald, Analyzing the acoustic communication of dolphins, in *Dolphin Communication and Cognition: Past, Present, and Future* (2015), p. 107
5. P.L. Tyack, C.W. Clark, Communication and acoustic behavior of dolphins and whales, in *Hearing by Whales and Dolphins* (Springer, New York, NY, 2000), pp. 156–224
6. H. Uchida, M. Zhu, The universal networking language beyond machine translation, in *International Symposium on Language in Language and Cyberspace* (2001), muslimbi.com
7. G.A. Miller, R. Beckwith, C.D. Fellbaum, D. Gross, K. Miller, WordNet: an online lexical database. Int. J. Lexicograph. **3**(4), 235–244 (1990)
8. http://www.cfilt.iitb.ac.in/wordnet/webhwn/wn.php
9. https://www.everything2.com/index.pl?node=flies

Chapter 4
Deep and Minimal Parsing

4.1 Introduction

We saw that two approaches are possible for dealing with user queries in an HMI system, namely,

1. Keyword-based approach, i.e. using only keywords without attention to grammar: This might result in the machine not being able to identify the intent of the query and hence stacking up a large number of largely irrelevant responses which are selected by the identified keywords in the query or
2. Deep parsing approach: Here, the system may end up being unable to parse the query because of grammatical inconsistencies in the query, it may, on the other hand, accept sentences or queries like
   - `The rude sky ate the singing chair` or,
   - `Who was the 2016 of head during CompanyX`

Deep parsing is generally thought to be necessary to extract the syntactic, hence the semantic relationships between the words. However, parsing sentences are quite difficult for a machine to do. UNL type of analysing sentences resolves these inconsistencies, but requires human involvement, working in a semi-automatic mode.

The question is whether, instead of these extremes, can one think of a way of extracting such a structure without formally parsing the sentence. We can pose this question because we know for sure that humans can do this. For instance, they can deal with ungrammatical and even incomplete sentences which we normally use in day to day natural conversation, though these sentences are impossible to parse. Second, illiterate persons and young children can understand sentences even though they do not know grammar or may have not been exposed to formal grammar.

Our contention is that there is a middle path or approach, called *minimal parsing* that should be able to emulate what the human does: make sense of queries which are incomplete or grammatically incorrect or both. We believe that this kind of

© Springer Nature Singapore Pte Ltd. 2018

P. V. S. Rao and S. K. Kopparapu, *Friendly Interfaces Between Humans and Machines*, https://doi.org/10.1007/978-981-13-1750-7_4

in-between approach should be useful for enabling user-friendly natural language human–machine interfaces.

We also maintain that while UNL type of complexity and comprehensiveness are neither necessary nor desirable, a simpler WordNet type of awareness of meanings of words and their mutual relationships is needed. Additionally, rather than deal with syntax and semantics separately and in isolation with each other, it would be best if both these are dealt with together in an integrated manner. The same principles are valid in most routine day to day conversations between individuals. Even in normal conversational language, sentences are often not grammatical; quite frequently, they are not even complete. This is because, as seen earlier, syntactic constraints are not crucial for query understanding.

It should be noted that syntactic and semantic constraints are necessary for answer generation or synthesis; otherwise, the generated sentences will not be grammatical (however they may still make sense—as in day to day conversations). Syntax, however, may help to resolve ambiguity if any.

It might be argued however that this may not always be possible. For example, consider the following sentences.

- `The president whom everyone respects resigned.`
- `Who did you say came here?` (Non-local dependencies)
- `The boy who generally stood first in class failed this time due to diversions such as cricket where less than a week back, he scored a century`" (Complex structure).

In such cases, systematic parsing is needed. We can quickly see that the semi-literate person too would have difficulties with such sentences. Even for the well-educated listener, it is in general difficult to deal with such sentences quickly in a spoken interaction, without actually '*seeing*' the sentence, say on paper. This involves too much cognitive effort and hampers free communication. That is why, in normal conversation, we often need to repeat complex sentences or even rephrase them.

It follows that **conversational sentences should not be complex**. **So long as they are amenable to easy understanding** without complex processing **and the intended meaning is extractable on the fly, they need not even be grammatical. This is the type of task domain that we wish to address.**

Thus we see that there should in general be no need to use complex grammars and complicated word processing in human interaction. It follows that rigorous syntactic analysis is <u>not crucial</u> for understanding the intent. In general, we <u>do not</u> check whether grammar rules are being followed in the sentence.

Our scheme of using what we call a keyconcept-keyword (KC-KW) or minimal parsing is effective in accomplishing this in most cases. To understand the idea behind the notion of "keyconcepts", let us examine whether in each of the sentences below, there exists *one word* (key constituent) which determines the nature of these semantic relationships?

1. `Ram slept.`
2. `Ram saw Sita.`
3. `Ram gave Sita a book.`

Transaction	# constituents	first	second	third	fourth	fifth	sixth	seventh
Sleep	1	subject						
See	2	subject	object seen					
Give	3	subject	person receiving	item giver				
Purchase	variable	purchaser	object purchased	seller	price	rate	quality	amount
Movement	variable	destination	origin	means	time	date	route	

Fig. 4.1 Transaction and number of constituents

4. `Ram purchased chocolates for Sita.`
   a. `at the market`
   b. `at the market for INR 200.`
   c. `at the market for INR 200 per kilogram.`
5. `I want to go`
   a. `to Delhi`
   b. `from Kanpur to Delhi`
   c. `from Kanpur to Delhi by air`
   d. `from Kanpur to Delhi by air on April 5`

Clearly, there are patterns in these sentences. The number of quanta of information that each sentence contains by and large depends on the transaction (which is usually the verb in a sentence); so, if you know what the transaction is, you know what constituents to expect. This word which indicates the transaction is called the keyconcept or head constituent.

As seen in Fig. 4.1, if the transaction is "`sleep`", then it would require just one constituent (noun, living being) which is the subject (e.g.: `Ram slept`) and a transaction "`see`" would require two constituents "a subject" (noun, living being) and "an object" to be seen (noun, living or nonliving): (example: "`Ram saw Sita`" or "`Ram saw a book`"). Similarly, the transaction of "`give`" would need three constituents "a subject", "a receiver" and the "item" (example: "`Ram gave Sita a book`" or "`Ram gave 100 rupees to an NGO`").

In every sentence or phrase (as in the above examples), one word is dominant, i.e. associated with independent features. This is the "head constituent". The features of other words (other than the head constituent word) depend on the head constituent word. For example, each of this sample phrase has a head constituent which is underlined. The head constituent can be said to be not dependent on anything else in the phrase or a sentence, while the other words in the sentence are dependent on the head constituent; each of the constituents has to satisfy two criteria, namely,

1. they have to belong to the correct **syntactic** category, and
2. they have to belong to the correct **semantic** category

The verb is the most important word in a phrase and could have zero, one or more arguments or constituents. For example,

1. Zero arguments **rains()**

   a. `it` **`rains`** (V)

2. One argument **sleep(subject)**

   a. `Ram` **`sleeps`** (V)

3. Two arguments **talks_about | loves(subject, object)**

   a. `John` **`loves`** `Mary` (V, NP)
   b. `John` **`talks about`** `philosophy` (V, PP)

4. Three arguments **gives(subject, receiver, object)**

   a. `John` **`gives`** `Mary a book` (V, NP, NP)
   b. `John` **`gives`** `a book` **`to`** `Mary` (V, NP, PP)

5. Four arguments **purchase(subject, receiver, object, price)**

   a. `John` **`purchases`** `a book from Mary for INR 50/-`

It can get quite complex with transactions like "`purchase`" and "`movement`" which can have a variable number of constituents. For example,

- purchase (3 constituents): `Ram purchased chocolates for Sita`
- purchase (4 constituents): `Ram purchased chocolates for Sita at the market`
- purchase (5 constituents): `Ram purchased chocolates for Sita at the market for INR 200,`
- purchase (6 constituents): `Ram purchased chocolates for Sita at the market for INR 200 per kilogram,`
- purchase (7 constituents): `Ram purchased chocolates for Sita at the market for INR 200 per kilogram yesterday.`

To re-emphasize what we stated before, most day-to-day interaction between humans happens without strict and deep grammatical analysis of the sentences: for example, conversations between illiterate or semi-literate individuals. This is also true in conversations pertaining to specific task domains (e.g. ticket reservation). In fact in such situations, communication falters—even breaks down—if one uses complex sentences requiring detailed analysis of deep structure; such complex sentences are usually used in intellectual discussions or discourses involving learned participants, where subtlety of expression and precision of meaning are paramount.

Keyword-based simple sentence/phrase based communication seems to happen in the following types of contexts:

1. Highly repetitive tasks (e.g. ticket booking at railway stations and movie theatres)
2. Language disability (new language in a foreign country)
3. Coded communication (e.g. in front of children)

To elaborate, consider a human at a railway station purchasing a ticket. If the request is

`Give me a second class ticket by Deccan Queen to Pune for tomorrow.`

The booking clerk presumably ignores all words other than "`second`", "`Deccan Queen`", "`tomorrow`", "`Pune`". In fact, the human could have just said

`second class tomorrow Deccan Queen Pune`

Here, since the transaction is at a ticket counter, the keyconcept ('ticket purchase' is implied). Based on that keyconcept or head constituent and the other constituents above, the passenger would still have been able to make the ticket purchase. Clearly, the keyconcept has to be explicitly stated if the counter caters to **several types of requests.**

While this was a simple example, we could look at a slightly more complex situation in the scenario of railway enquiry.

`could you tell me if there is an air conditioned train from Delhi to Chennai via Nagpur on Monday?`

Observe that even here, as in the previous example, the information consists of the head constituent and several other *constituents.* Each constituent comes as a word or phrase; each of these can be substituted with other words or phrases and request for action would still be a valid. For example,

- `air conditioned` could be replaced by `nonstop` | `express`
- `Monday` could be replaced by `tonight` | `next month` | `in summer`

Here,

- information comes in discrete **'quanta'.**
- Each quantum is a phrase/word.
- Number of quanta depends on situation
- Even for a given situation this number may vary.
- Each quantum occupies a slot whose possible entries belong to (a) the same syntactic category and (b) have similar semantic sense. All permitted entries for a slot can be said to belong to the same *syntacto-semantic* (SS) category.

For example, consider the following sentences:

- `Could you tell me if there is an air conditioned train`
- `Could you tell me if there is an air conditioned express from Delhi`
- `Could you tell me if there is an sleeper train from Kolkata to Chennai`
- `Could you tell me if there is a nonstop train from Mumbai to Allahabad via Nagpur`

- Could you tell me if there is any type of train from Delhi to Chennai via Nagpur on Monday?
- Could you tell me if there is an air conditioned train from Delhi to Chennai via Nagpur on August 12th in the morning?

We can see that the number of quanta of information keeps progressively increasing in the above sequence of examples. Also, a given slot for a quantum of information is occupied by entries belonging to the same syntacto-semantic (SS) category. For instance, air conditioned train, sleeper train, nonstop train belong to the SS category noun phrase, train type. The entries Mumbai, Kolkata, Nagpur, etc. are noun, city name. Monday, August 12th belongs to noun, day/date. Now let us look at another example—a Hindi construct:

Introductory	krupaiyaa bataaiye ki (please tell) Aap (You)
Time frame specification	aath janvary ke baad (After 8th January) aath tareekh tak (Up to the eighth instant)
Source (Place name) specification	haidrabad say (From Hyderabad) haidrabad dwara (via Hyderabad)
Destination (Place name) specification	haidrabad jaanewaali (Going to Hyderabad) haidrabad kaa (For Hyderabad)
Time frame specification	aaj kaa (Today's) aath mahine ke bad (After eight months)
Train specification	andhra express may (In Andhra Express) kiseebhi gaadiyon may (In any trains)
Requirement specification (Broad)	aath seat (Eight seats) kya class ka (For which class)
Requirement specification (Fine) with end marker	confirm seat ho saktaa hai (Can the seat be confirmed) gaadi hai (Is there a train?)
End marker (Question)	kya (Is it)

Consider the following sentence

I ordered a book from Flipkart for INR 250 today

In this case, **'ordered'** is the keyconcept. Let us consider a few other sentences.

I shall be travelling to Delhi from Mumbai by Train on Sunday at 9 am

John ate an apple

Oswald killed Kennedy

Who killed Kennedy?

When does the train to Delhi depart?

In each of these sentences, this crucial word or keyconcept is underlined.

The first sentence (`I ordered a book from Flipkart for INR 250 today`) can be rewritten as

`Order (I, book, Flipkart, INR 250, today)`

Note that it is easy to understand the meaning even in this form.

`Travel (I, Delhi, Mumbai, train, Sunday, 9 am)`

Now let us look at the sentence

`Oswald killed Kennedy`

Which can be rewritten as

`Kill (Oswald, Kennedy)`

Now look at the question

`Who killed Kennedy?`

and rewrite it as

`Kill (? Kennedy)`

For the sentence

`Eat (John, apple)`

We can have

`Eat (?1, apple)` or

`Eat (John, ?2)`

In this sentence, we know (using associated knowledge, the dimensionality or syntacto-semantic category needed floor the keyconcept `eat`) that the answer `?1` has to be a living organism (noun, living organism). Similarly the answer `?2` has to be something edible (noun, edible substance). For example, then.

`eat (sky, table)`

is not permitted.

Let us illustrate this concept with a few more examples from another angle. Consider the mathematical formula for area of a rectangle.

Area = Length × Breadth

Area (number dimensionality L^2) = Length (number dimensionality L) × Breadth (number dimensionality L)

Distance covered by a moving object (Newton's law)

$s = u\,t + 1/2\,a\,t^2$

Distance covered s (number dimensionality L) =
Initial Velocity u (number dimensionality LT^{-1}) × time t (number dimensionality T) + 1/2 × acceleration a (number dimensionality LT^{-2}) × time^2 t^2 (number dimensionality T^2)

These formulae specify the nature and dimensionality (syntacto-semantic properties) of the individual parameters appearing in it. Observe that "u t" and "1/2 a t^2" have the SS property of "L" which is the SS property of "s" the distance covered.

From formulae like this, one can answer questions like

- `What is the area, if length is 2 meters and breadth is 1 meter?`
- `What is the length, if area is 6 square meters and length 3 meters?`

This is because the nature and dimensionality of the constants and variables conform to the specifications in the formula.

However, a question like

- `What is the breadth if area is six meters and length is 3 seconds?`

is invalid because the slots do not have valid SS category or dimensionalities. In this case, area cannot have the dimension of "L" and length cannot have the dimension of "T".

4.2 Syntacto-Semantic (SS) Categories for Phrases

Consider the query sentence

> "`Who was the person you were roaming with yesterday at 5.00 pm near the market`"

We can break this into syntacto-semantic categories as in the following table which shows the syntactic type or part of speech (PoS) and the semantic type (Type) of each word or phrase.

Phrase	PoS	Type
`Who was the person (?)`	Noun	Person's name
`you`	(Pro)Noun	Person's name
`were roaming with`	**Verb**	**Activity**
`yesterday`	Noun	Day
`at 5.00 pm`	Phrase	Time
`near the market`	Phrase	Location

For the next example, the keyconcept is **join.**

`How do I join XYZ Infotech?`

Phrase	PoS	Type
How do (?)	Verb	Process
I	(Pro)Noun	Person name
join	**Activity**	**Activity, Transformation**
XYZ Infotech	Noun	Organization

`What is the price of 1 tonne of cement?`

Phrase	PoS	Type
What is the (?)	Phrase	Interrogative
Price	**Noun**	**Cost**
of 1 tonne	Noun	Quantity Weight
of cement	Noun	Commodity

`I am a Physics graduate with experience in Computer Programming`

Phrase	PoS	Type
I am a	**Phrase**	**Assertion**
Physics	Noun	Subject
Graduate with experience	Noun	Qualification
In Computer Programming	Noun	Specialization

By understanding the syntacto-semantic structure of the word or the phrase in the question one can make sure that the answer provided is of the type expected. In the examples shown above, for the query

“Who was the person you were roaming with yesterday at 5.00 pm near the market”

it is expected that the answer is the name of the person.

For the query

“How do I join XYZ Infotech?”

The answer has to be a process description.

Note that in all the above examples, the keyconcept word ‘holds’ or ‘binds’ all the other keywords together. More importantly, if the keyconcept word is removed, all the other words fall apart.

4.3 Minimal Parsing System [1]

There are several ways in which keyconcepts can be visualized, we enumerate a few,

1. Keyconcept can be visualized as a mathematical functional which links other keywords to itself. Informally keyconcepts are broadly like 'mathematical functions' which carry 'arguments' with them, for example, KC1 (KW1, KW2, KC2 (KW3, KW4). Here KC1 and KC2 are the two keyconcepts and KW1, KW2, KW3 and KW4 are the keywords. Note that KC2(KW3, KW4) is a function with KC2 binding the dimensionality of the arguments KW3 and KW4 as mentioned earlier and KC1 (KW1, KW2, KC2 (KW3, KW4) is a functional (function of a function) which binds together KW1, KW2 and KC2(KW3, KW4). Given the keyconcept (in our example KC1 and KC2), the nature and dimensionality of the associated keywords (KW1, KW2 and KW3, KW4, respectively) get specified. We define the arguments in terms of syntacto-semantic variables: For example, destination has to be a noun which is a valid city name; the price has to be a noun number corresponding to a currency etc. Some examples of keyconcepts determining the syntacto-semantic nature of the keywords are (a) Mass(density, volume(length, breadth, thickness)) or Mass (density, volume(radius)); (b) Purchase(purchaser, object, seller, price, time); Travel(traveller, origin, destination, mode(class_of_travel), day, time).
2. Keyconcept can be also seen as a template specification: if the keyconcept is purchase/sell, the keywords will be material, quantity, rate, discount, supplier, etc. Valence (as in Chemistry) or the number of arguments that the template determined by the keyconcept is known once the keyconcept is identified.
3. Keyconcept can also be seen as a computer database structure specification: consider the sentence, `John travels on July 20th at seven pm by train to Delhi`. The underlying database structure would be

KC	KW1	KW2	KW3	KW4	KW5
Travel	Traveller	Destination	Mode	Day	Time
	`John`	`Delhi`	`Train`	`July_20`	`seven pm`

Keyconcepts together with all the KWs not only help in capturing the total intent of the query but also enable gauge the semantic validity of the query. As a consequence, the query can be interpreted in a very specific manner and the search in the answer space is constrained. For this reason, an HMI system based on keyconcept-keyword (or minimal parsing) can be a very effective, practical and usable solution to the problem of natural (spoken) language understanding in a wide variety of situations.

The introduction of keyconcept (KC) gives KC-KW based system a significant edge over keyword (KW) only question answering or question understanding systems. Identifying KCs not only helps in establishing the semantic validity of the query but also results in better understanding of the query and hence the KC-KW based system is able to answer the query more appropriately in a way a human would do. As we will see later this gives the minimal parsing system the unique ability to give a "feel" answer rather than just a logical answer. While one can construct a meaningful query with several KC's, it should be noted that a typical meaningful transactional query, in all likelihood, will consist of only one KC.

Note that in general a paragraph of information, considered for the purposes of understanding and analysis, rides on a single KC. However, in the event of more than one KC being present, one can treat them as a hierarchy of KC's.

A typical working of a minimal parsing (KC-KW) system would be

1. Look at a query Q = {w1, w2, w3, w4, … wn} being made up of say *n* words
2. Remove stop words (connection words, most commonly occurring, etc.) from Q to get Q′. Note that Q′ is a subset of Q.
3. Minimal parse Q′ to obtain mpQ′ = KC(KW1, KW2, ….). Note that KC, KW1, KW2, etc. are words (or phrases) in Q′
4. Look for and match appropriate answer paragraphs which have also been minimal parsed and represented in the KC-KW form.
5. Exact answers paragraphs are those which have identical pattern as that of Q′ (or Q)
6. In case there is no answer paragraph which exactly matches the query, then,
   a. Generate mpQ″ = KC(KW1′, KW2, …) from mpQ′ using a taxonomy tree of the type we saw for WordNet; e.g. Synonyms, hypernyms or siblings (for example "snack" is a hypernym of "dosa" and "idli" is a sibling of "dosa") such that kw1′ is *taxonomically close* to kw1.
   b. Look for and match appropriate answer paragraphs which have also been minimal parsed to pattern match mpQ″
   c. Then generate mpQ‴ = KC(KW1, KW2′, …) from mpQ′ using a taxonomy tree such that KW2′ is taxonomically close to KW2.
   d. Look for and match appropriate answer paragraphs which have also been minimal parsed to pattern match mpQ″
   e. Repeat this for other kw's and kc
7. If Step 6 does not result in a matched answer paragraph. The query Q does not have an answer in its knowledge-base. Else, we get an answer paragraph with approximate match which can result in (either a more generalized or a nearest available) response.

We will show several examples of how these approximate matches are identified through examples in a later chapter which describe some of our case studies.

For question understanding, we can go forward even if the rules of syntax are not followed in sentence construction. This means that we can ignore these rules during analysis. For example, consider the sentence,

```
Dashrath gave Ram a book
```

Here, apart from word order, the fact that a book, being an inanimate object, can be the 'object' given. Ram, being a person can be the receiving agent. In other words, "give" is the keyconcept having three keywords (Dashrath and Ram—noun, person; book—noun, object). In this type of analysis without full parsing, there would be ambiguity between Ram and Dashrath about who is the giver and who the taker. So both alternative interpretations would be taken into account.

Though this is not a constraint, we will assume that each answer paragraph has only one KC. Note that if a single answer paragraph has multiple KCs, then we can segment the answer paragraph into multiple smaller answer paragraph such that each answer paragraph has one KC. One can think of a QA system based on KC and KW (or minimal parsing) as one that would save the need to fully parse the query. As mentioned earlier there are many constraints on the query for full parsing to work accurately and more often than not these constraints are not valid on transactional queries. However minimal parsing comes at a cost; use of this approach could result in the system not being able to distinguish who killed whom in the sentence

```
Upen killed Mandar
```

The minimal parsing system would represent it as

```
kill(Upen, Mandar)
```

This minimal parsing representation can have two different interpretations, namely, Upen killed Mandar or Mandar killed Upen and it is not possible to interpret which of this two is true from kill(Upen, Mandar). But in general, this is not a serious issue unless there are two different paragraphs: the first paragraph describing "Upen killing Mandar" and a second paragraph (very unlikely) describing "Mandar killing Upen". If indeed two such paragraphs exist, there could be an ambiguity of which answer to present but note that such confusions can and do occur even in human–human communication.

It may be mentioned in this context that given the question

```
"How many rupees to the dollar",
```

Google [2] responds

```
"1 Indian Rupee equals 0.016 US Dollar"
```

It is easy to see that this is not the required answer but the *reverse* of it. This shows the same kind of confusion that is likely to happen when we use the minimal parsing approach.

There are reasons to believe that humans resort to a keyconcept type of approach in processing word strings or sentences exchanged in bilateral, oral interactions of a transactional type. As mentioned in [1], a clerk sitting at an enquiry counter at a railway station does not carefully parse the queries that passengers ask him. That is how he is able to deal with incomplete and ungrammatical queries. In fact, he would have some difficulty in dealing with long and complex sentences even if they are grammatical.

As we saw, once the keyconcept is identified in a query, one also immediately knows what other types of keywords to expect in the query; the relevant keywords in the query can then be extracted—based on the syntacto-semantic categories of the concerned words. **Transactions are often easier to perform if the keyconcept is spoken first, then the keywords.** Reversing this order can lead to difficulty in comprehension.

4.4 Using the Minimal Parsing Idea

The primary goal of most question answering (QA) systems is that they be usable or easy to use by humans. From the input perspective, the system should not impose any constraints on the way a person can pose queries, e.g. having to construct syntactically correct queries. From the output perspective, however, it would be highly desirable (though not strictly essential) for the system to give out grammatical responses. It would be important for the system to fetch a correct or, in case that is not available, an approximate response to a query.

The middle path approach, namely, minimal parsing should work well in this kind of a scenario by identifying two types of parameters, namely, keyconcepts (KC) first and then the associated keywords (KW).

To further clarify syntacto-semantic categories, let us look at a railway reservation inquiry related to a ticket transaction. While in this scenario, the main constituent (keyconcept) would be a ticket transaction category (a verb) like `purchase` | `confirm` | `cancel` this main constituent would determine the syntactic and semantic categories of other constituents (keywords/phrases). For example, a purchase of a ticket will need to have the name of the city, the name of the train and the class of travel. So the three constituents (keywords) that are attached to the main constituent (keyconcept) need to belong to certain SS categories.

1. noun—city name
2. noun—train name
3. noun—seat category

Given the above, minimal parsing of a sentence or a phrase would require two important steps, in a certain order, namely,

1. identification of the syntacto-semantic category of the main constituent (keyconcept) and
2. based on that, identification of the phrases containing the other constituents, based on their expected SS categories.

For example,

- `in the morning,`
- `after 3 pm,`
- `between April and July,`
- `from Bombay,`
- `to Delhi,`
- `by the Rajdhani Express`

4.5 KC in Transaction Oriented HMI

An HMI system based on keyconcept and keyword approach can be an effective practical solution to address natural (spoken) language understanding in a wide variety of situations. In the following section, we briefly describe some of the systems which have been implemented using this approach.

Most interactions between a service provider (or business) and a customer tend to be transaction oriented. The customer of an organization usually approaches a business for performing a transaction or enquiring to get very specific information. For example, it could be to get information about

- balance in their savings bank account,
- status of their address change request
- advance booking for cinema tickets
- purchase of train, bus, air tickets and so on.

Even actions like transfer of money, purchase and sale of debentures and stocks could also proceed similarly (provided security screening, in the form of person authentication, has already taken place).

(1) Conversation between the customer and the service provider is generally restricted and does not go beyond the specific task on hand. Note that a customer does not discuss politics or weather with the booking clerk at a railway counter.
(2) Sentences used, during the conversation, are simple to understand and straightforward to avoid any ambiguity. Often, the conversation is not even in complete sentences. Rarely is any parsing required and not every single word is attended to.
(3) The only need is to elicit the intended meaning, with minimal effort
    (a) scan utterance to determine the keyconcept
    (b) find out relevant keywords (belonging to the desired SS category).
(4) This works even with phrases and ill-constructed, or unfinished sentences.

As we mentioned earlier, keyconcepts are usually (but not necessarily) verbs. Examples are `reserve`, `activity`, `plan`, `manage`, etc. Examples of non-verb like keyconcepts that can be found in transactional interactions could be `availability`, `concession`, `senior citizens`, etc.

There can be a few possible areas for ambiguity because of the KC-KW representation due to minimal parsing. For instance, as we saw earlier, for the key-concept: `kill`, the keywords would be "killer", "victim", "time", "place", etc. Confusion can arise between "killer" and "victim" because both have the same SS category of dimension (namely, name of human), e.g.

```
kill who Oswald?
```

Can be interpreted as

(a) "`who did Oswald kill`" (which should result in the answer "Kennedy"), or

(b) "`who killed Oswald?`" (which should result in the answer "Jack Ruby").

Both interpretations would be correct; and subsequently, it may not be very difficult to find the required answer to a question if it is available in the knowledge-base. Note that we have already seen that such confusion occurs in Google answers.

The more important question is what happens if the answer is not in the knowledge-base. There are two options:

1. Give up, indicate failure
2. Give next best usable information

For the question "`Do you have Colgate toothpowder?`", one can give one of the following two answers:

```
1. No I'm sorry.
2. I have Colgate toothpaste, will that do?
```

While both the answers are correct, the second answer is more useful and is very much preferred. In order to be able to retrieve not just the exact information required (it may not even exist in the knowledge-base), it would be desirable to be able to perform a *broader* search and locate even approximations to what one wants. This ability requires knowledge of **relationships** between words (meanings) or the ability to find semantic approximations.

While this might look like an intractable problem—this can be attempted by writing out ad hoc lookup tables which capture the semantic relationships. However, it would be useful to have a formal framework which can form the basis for any specific method of implementation that the designer of a QA system might adopt. To make this possible, we can use a domain-specific taxonomy tree along the lines of WordNet (that we described earlier) which is one such formal framework. See Appendix A.1 (Example Taxonomy Tree) for an example of a handcrafted taxonomy tree built by us for a book on Information technology.

References

1. S. Kopparapu, A. Srivastava, P. Rao, Minimal parsing question answering system, in *International Conference on HCI*, PR of China, 2007
2. Google, in http://www.google.com, website

Chapter 5
HMI Systems

With advances in technology in varied fronts and growth in the services and allied industries, it has become quite common for humans to interact with machines on a frequent basis. In contrast with question answering systems (where the interaction is generally supposed to be *one-shot*), there is need for chatbots (text input) and voice-bots (voice input). Chatbots are dialogue systems where the human and the machine interact taking turns; each turn could be loosely looked upon as a Question Answering in an HMI engagement. A sequence of QA engagements could be looked upon as an interactive chatbot session.

It is the endeavour for any researcher working in the broad area of artificial intelligence (AI) to make this interaction as seamless, natural and appear as if the interaction is actually between humans. A natural language interface is the most effective interface between a human and a machine; a good natural language interface breaks—or at least narrows the interaction barrier between human and machine, and makes it look more like human–human interaction.

Humans communicate with each other for a variety of reasons. This communication/interaction can be of several types, for example:

1. Questions: `Did it rain yesterday?`
2. Assertions: `He went home.`
3. Requests/Directions: `Please pass me the salt; Can you tell me what time it is?`
4. Reactions: `I am happy to hear that!`
5. Expressive Evaluations: `He is a very considerate person.`
6. Commitments: `I will repay you tomorrow.`
7. Declarations: `India is a sovereign, democratic, secular republic.`

We have already emphasized that there is an increasing need for communication between people and machines. Keeping aside the question of whether in the near future all types of human–human interaction might be required to be implemented in human–machine interactions, we can see that there exists a need for at least four types

© Springer Nature Singapore Pte Ltd. 2018

P. V. S. Rao and S. K. Kopparapu, *Friendly Interfaces Between Humans and Machines*, https://doi.org/10.1007/978-981-13-1750-7_5

of interaction, namely, (1) Questions (seeking information), (2) Requests/Directions (for specific actions), (3) Assertions (regarding facts) and (4) Declarations (of universal truths). And of these, the first two are high priority in terms of need.

Question answering systems (QASs) are a special case of HMIS's; these are, typically, machine-driven automatic systems that can respond to queries in natural language posed by humans. There are several flavours of QAS's and these are ubiquitous in today's increasingly digital landscape. Question Answering systems manifest themselves in different forms. At one end of the spectrum one could visualize the simplistic "search box option" common on most websites to be a QAS; here the search words are the questions and the returned hyperlinks or pages are the machine-generated answers. And at the other end of the spectrum, one could consider the more recent chatbots as a form of QAS. Here a human and machine interact, perhaps over several exchanges of remarks, the final goal being to address the needs that the human might have.

The need for QA's is easily seen. Even the really well designed and organized websites are not easy to traverse, specifically if one has a very particular specialized need. Studies have shown that individuals give up easily if websites are not user friendly and do not quickly provide the specific information they are seeking. In a competitive economic world, this can mean loss of business. According to a study [1], the hit rate (ratio of number of completed transactions to number of online hits) is less than 3%. Manual customer interaction services and customer support representatives are highly expensive, making the need for question answering systems very pressing.

A QA system should

1. accept inputs in natural language
2. extract the intent of the input query (not necessarily by parsing it)
3. extract the required response from the (textual or conventional) knowledge-base.

Any user-friendly natural language interface would need essentially two blocks, the first block would need to interface with the human and act as a transducer to convert the audio or text from the human into a computer compatible string; for our purposes, we can assume that speech recognition has been done (if the input was speech, as is in the case of a voice-bot) with a good degree of accuracy (else one could resort to correction of speech recognition output as mentioned in [2]) and that the text that we need to deal with is a computer attuned string. The second significant block is the intelligent block that "understands" the query from the human. This is the most important block in a user-friendly natural language interface.

Figure 5.1 is a broad schematic of how a QA system works. Here Q is the query from a human in a natural language and A is the response provided by the QA system. The intent is extracted by the question analyzer block which in turn produce machine processable queries. These queries help in information or document retrieval. The last block allows for assembling together the information in the retrieved documents in a human digestible form. Most QA systems stop at extracting the information or at the document retrieval stage. In this monograph, we concentrate on analyzing the query and retrieving information (answers). The answers are assumed to be well articulated and hence readable and informative.

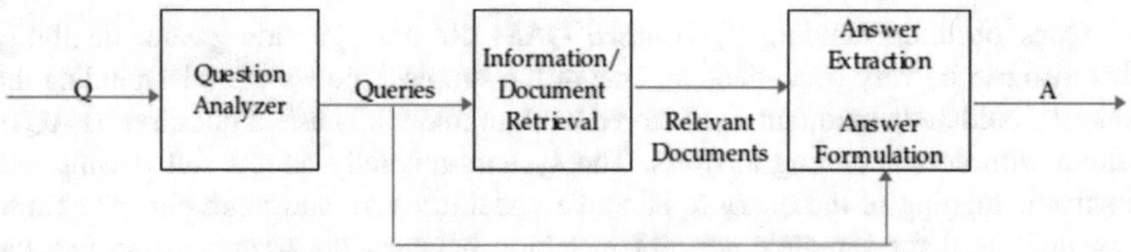

Fig. 5.1 QA system schematic

Some of the well-known methods used in QASs are

1. Keyword and Tri-gram (shallow language understanding)
2. Statistical techniques
3. Expert systems
4. Question understanding

In a keyword and trigram (shallow language understanding) method, the system tries to match words in the human query with similar words in its answer or knowledge-base either as individual words or word triplets (3-gram) or a combination. Since these systems match lexical words alone without any specific attention to semantics, they perform very poorly in general. Performance of such systems improve when certain lexical words (example keywords) are prioritized by assigning them more weightage depending on the context or the domain. Some QA systems meta-tag the keywords as required, optional, irrelevant and forbidden to improve the extraction of the answers. These systems are unable to distinguish between the "how" and "why" type of questions. For example, "`How can I open a savings account with you`" and "`Why should I open a savings bank account with you`" would return the same set of answers. Usually, such systems discard the 'wh' word in the question. Such systems might also be designed to reject any answer that contains too many non-envisaged keywords. These small tweaks make keyword-based QASs reasonably "smart".

In statistical methods, one uses artificial neural networks (ANN) and Bayesian approaches. The more powerful statistical methods match the query word with attributes. For example, "`cat`" can be equated to "`furry animal`", "`tail`", "`drink milk`", "`whiskers`", etc. which leads to a wider scope but at the same time this could lead to false leads and there is a tradeoff. As can be imagined, statistical methods have to be set up separately for each domain application and they generally tend to fail if there are new data or information additions to an existing system or when faced with a new domain.

Expert systems provide for rule-based inference which is highly intensive in effort when rules are being formed or built. As one can expect, the rules must be either manually written or automatically constructed using machine learning (ML) techniques for each domain. Once the rules have been handcrafted, updating them to suit new data is quite effort intensive.

Question understanding (QU) based QASs not only provide greater flexibility but also can be very cost effective. One of the simplest modes of understanding the query would be to maintain a pre-stored bank of frequently asked questions (FAQs), along with corresponding answers. The system generally adopts full parsing and syntactic tagging of the query to effectively identify verb and noun phrases. These are then used for semantic concept matching between the user question and the FAQ entries (which have answers tagged to them). The pre-stored answer to the best matching question is given out as the response of the system. For domain-specific scenarios which usually have a small set of "types of questions", even partial understanding can yield reasonably satisfactory responses.

To illustrate, let

$$\text{Qfaq} = \{\text{q1}, \text{q2}, \ldots \text{q}m\}$$

be the set of most frequently asked queries and let {q1, q2, … q*n*} (*n* is less than or equal to *m*) be unique queries, abstracted from these and let {a1, a2, … a*n*} be the answers to the queries Qfaq′ = {q1, q2, … q*n*} respectively.

The question understanding system would take the human question Qh and compare it to all the existing queries in Qfaq′ and find a question (or a set of questions in Qfaq′) which is (are) closest (close) to Qh.

Let us say,

$$\text{qk}^* = \max_{\text{qk}\,\varepsilon\,\text{Qfaq}} P(\text{qk}|\text{Qh})$$

matches the human query Qh best. The system would then respond to Qh with an answer ah* corresponding to the qk* in Qfaq′. Depending on the type of application or the context one could rank order P(qk | Qh) for all qi ε Qfaq′ and choose the top few. For example, if P(qa | Qh) > P(qb | Qh) > P(qc | Qh)—then the system responds with "aa" and "ab", corresponding to the "qa" and "qb" respectively, as the top two responses to the user query Qh.

Systems adopting this technique work best for queries like

- "`how can I open a saving account`"
- "`whom should I contact to report my lost credit card`"
- "`How do I change my mailing address?`"
- "`How do I transfer money into my account?`"

However, questions that one might have to deal within a question answering system, like

- `Show me all display devices which are LED and between 8 and 12 inches in size.`
- `Show me all IT stocks costing less than Rs.200/- per share.`
- `Which of these give more than 15% returns?`

cannot be handled because such systems will not be able to deal with the *more than*, *less than* and *between* type of questions. A more sophisticated understanding of the question is desired to answer such queries.

Taxonomies of QA systems have been nicely captured in [3] in the form of five different classes. As can be seen, there are five different classes of QA systems. There are mostly based on the type of knowledge-base that the QA systems need to address. At one end is the use of simple knowledge sources which are able to address questions like "`What is the largest city in India?`" systems of this kind make use of simple semantics and are usually based on indexing paragraphs of information. At the other extreme is when a QA system has to answer based on the world knowledge, these systems are generally special purpose and dwell into deeper level of language processing involving techniques in text mining, dialogue and meaning negation. These special purpose QA systems are able to answer questions like "What should India's policy be towards Israel?". Most QA systems that are useful for transaction related queries on one hand do not need very sophisticated (Class 4, Class 5) QA system and on the other hand do not work for Class 1 type of QA systems. Subsequently, usable systems are the ones that are based on ontologies or taxonomies. These QA systems make use of the semantics of the words through taxonomies and are able to perform well for transactional type of QA systems.

Class	Knowledge-base	Reasoning	NLP/Indexing	Examples, Comments
1	Dictionaries	Pattern Matching, appositions, heuristics	Complex Nominals, Simple semantics, para Indexing	What is the largest city in India?, Calcutta, the largest city in India, …
2	Ontologies	Lightweight abductive reasoning	Morphological and semantic alterations of keywords	How was Mahatma Gandhi assassinated?, Mahatma Gandhi was shot by …
3	Very large KB framenet	Template based	Fusion of answers, coherent generation of answer	What are the arguments for and against prohibition?, Answer from many texts
4	Rapidly prototyped KBs (High-performance KB)	Analogical reasoning	Complex answers (compilation from diverse sources)	Will India lower the excise tax rates in this year? Answer contains analogies to past events
5	World knowledge	Expert level, special purpose	Text mining, dialogue, meaning negation	What should India's policy be towards Israel?, Answer is complex, referring to a developing scenario

Typical QA system can be enumerated by the processing complexity that is needed. As shown below information extraction is the easiest while the language generation is the most complex. An increasing order of complexity of process is captured in the list below. Most of these are very well known in natural language processing (NLP) literature or have been covered in this monograph elsewhere and for this reason, we do not elaborate on these here.

- Information Extraction
- Information Retrieval
- Syntax
- Semantics
- Pragmatics
- Reasoning
- Reference Resolution
- knowledge-base
- Language Generation

The list of technologies that enable QA systems are enumerated below.

- Part of speech (PoS) tagging is the process of assigning a grammar label (categories: verb, noun, etc.) to the words in a sentence. This allows for word category disambiguation based on both its definition and the context in which the word occurs.
- Parsing is a machine learning program that identifies the grammatical structure of a given sentence. Some of the tasks performed are (a) grouping words as "phrases" and (b) identifying which lexical words are the subject or object of a verb etc.
- Word Sense Disambiguation (WSD)—Generally, as mentioned earlier, words tend to have more than one sense (for instance "bank"). WSD is the problem of determining which "sense" (meaning) of a word is more appropriate in a particular context. Note that humans do it rather very well.
- Named entity Recognition (NER) is a subtask of information extraction that enables identifying and classifying named entities in a given text into predefined categories such as the names of persons, organizations, locations, etc.
- WordNet—We have covered this in detail in one of the earlier chapters. Needless to say, it is a handcrafted dictionary that is able to associate a relationship between words using semantics.
- Knowledge acquisition is the process of extracting, structuring and organizing knowledge, generally from one source.
- Knowledge classification enables knowledge reuse and sharing. Generally, one can classify the knowledge as being general, domain specific and enterprise specific, etc. For example, knowledge associated with banking domain probably can be used for all banking and insurance kind of domains; however knowledge specific to an enterprise (say xyz bank) might not be usable for another bank.
- Data base retrieval refers to information retrieval from a structured database. For example, a query like "Show me flights from Mumbai to Delhi" might require the generation of a SQL query so that this information can be extracted from a structured database. We will see some example in the chapter on case studies.

- Language Generation—The process of putting together chunks of text data to form a meaningful (semantically) and human readable (syntax) output.

Applications of HMI systems are several, we enumerate them briefly

- E-commerce—This is one of the most common area that requires an interaction between a human and machine. Typical questions that one might want to ask is "`Do you have detergent from a reputed brand that is being sold at a discount`" or "`Show me the flights from Mumbai to Delhi which fly via Jaipur`".
- Summarization requires significant amount of text processing. While most of the work in summarization in literature has been on news articles (which has a very specific pattern in terms of inverted information pyramid). Summarizing more complex documents, especially technical and legal document, requires significant amount of text processing.
- Tutoring | Learning—A QA system for an e-book which allows the student to ask specific questions and seek answers from the e-book quickly could be very useful. We elaborate on this in our case studies on an e-book on fitness (see Sect. 6.2).
- Personal assistant (Business, private) is another application that can benefit from HMI system. For example, a personal assistant to handle "`Set my alarm for ten minutes before 6 tomorrow`" to "`Please initiate a call to my supervisor immediately after my meeting at 4 pm`".
- Online trouble shooting is yet another application, typically a chatbot which allows the user the query a machine for a possible solution to fix computer not booting. For example,
  User: "`My machine is not booting what should I do`"
  Machine: "Did you check you if your machine is powered on"
  User: "`Yes I did`" ….
- Semantic Web—This is a process where in data in the web pages is structured and tagged in such a way that it can be read or parsed directly by computers. A fair degree of text understanding and processing is required to build robust text that can be automatically tagged to the web pages. For example, one could automatically create a meta-tag for a page using the content in the page to enable the web page to pop-up when a word in the meta-tag is part of the user query.

5.1 Information Retrieval Based Question Answering

Information retrieval (IR) based question answering systems [4] aim to furnish a response to the user query. The response is in the form of short texts (paragraphs of information) rather than a list of relevant documents which *might* contain the information that the user may be looking, for example, the output of a search engine (e.g. Google [5]) which returns hyperlinks to web pages | documents in response to a query).

These systems use techniques from computational linguistics, information retrieval and knowledge representation to identify the relevant answers.

Typical features of an IR based QA system are

- Question Analysis—Keyword extraction is the first step for identifying the input question type.
  - Identify Question Type—Often there are words that indicate the question-type directly, namely, "Who", "Where" or "How many". However, some words like "Which", "What" or "How" do not give clear picture of what should be expected answer-types, in this case, there is a severe restriction on the ability to gauge the expected answers types.
  - Word Meaning—Words in a query can have more than one sense. In situations like this,
    - Tagging,
    - Parsing (shallow | phrasal parsing) and
    - Template matching

    enable find the words that can indicate the semantic meaning of the question. A lexical dictionary such as WordNet can then be used for understanding the context which in turn enables imposing the correct sense to the word in the query.

    - Determine type of answer expected and build 'focus' for desired answer—Very often the question type imposes a restriction on the type of answer expected, namely, a question of the type "Who" expects the name of a "Person" in its answer, and a question of type "Where" would have to invariably contain a "Location" and a query like "How many" would need to have a "Number" in its answer.
    - Generate queries for search engine—using keywords, synonyms, hypernyms and hyponyms one can generate machine (e.g. SQL) queries which can be used to fetch answers from the knowledge-base (structured or unstructured).
    - Information extraction—From extracted documents, find the sentence(s) that best match the question. This does require a significant amount of language processing.
    - Additional
      - Name answer module to identify using heuristics words corresponding to persons and places, etc.
      - Numerical answer module units which can identify numerals corresponding to Length, area, time, currency, population, etc.

Some techniques and term definitions in IR

1. Information Retrieval could be based on some kind of pattern matching.
2. Coordinate matching where one makes sure that the dimension the answer is along the lines of the expected answer. As mentioned elsewhere in this monograph, an answer requiring Area (number dimensionality L^2) should match Length x Breadth each of number dimensionality L.
3. Stemming is the process of reducing inflected (or sometimes derived) words to their word stem. This is very useful in matching query and answer patterns.
4. Synonyms—allow the answers and question patterns to be matched from the semantic perspective rather than the exact string matching perspective
5. Identify important words—Keyword spotting to identify which word or words is to be given more importance, etc. Generally depends on the domain.
6. Proximity of question words—Finding a negate of a word might require analysis of a word preceding the keyword. For example, the word "not" in the query "`find me pictures that are not blue`" makes the answer type the opposite (in some sense) of the query "`find me pictures that are blue`".
7. Order of question words—The structure of almost every simple question is more or less based on the same model. For example, in English, the model is—Question word (if there is one) followed by auxiliary or modal word followed by the subject word followed by the main verb followed by the rest of the sentence. However, this order of this changes depending on the language.
8. Capitalizations—to extract words that are named entities, especially when answering 'Who" and "Where" type of questions, which expects the answers to contain a proper name of a person and location, respectively.
9. Quoted query words—might give some intended extra weight by the query seeker.

One of the main requirement though not explicitly stated is the need for both (a) the query and the (b) set of responses to be semantically and syntactically correct. This is dictated by the deep parsing that happens in most of the Information retrieval and question analysis modules as mentioned above. This requirement is very imposing on a human especially when he is using an HMI system for the purposes of transaction. As mentioned elsewhere in this monograph, and repeated here for completeness,

- the user has little time, and does not want to be constrained by how he can or can not ask for information
- the user may not be grammatically correct all the time (he would tend to use transactional grammar)
- a first time user is unlikely to be aware of the organization of the web pages
- the user roughly knows what he wants and would like to query in the same manner as he would query another human in his language.

5.2 Our Approach and Goal

Our goal is to implement a human-friendly HMI systems which understand the users input (in text) in natural language. These systems should tolerate even incomplete and nongrammatical sentences. Gives responses which as far as possible match the exact requirement of the user. A system should, as far as possible, not give up or fail; where an exact match is not possible, it should provide a best possible response (either a more generalized or a semantically nearest available) response. This means that we need a text processing approach that does not depend on rigorous parsing. It should take care of not just syntactic but syntactic as well as semantic relations between the words in the user input. Additionally, the system should

- be configurable to work with input in different languages
- provide information that is close to that being sought in the absence of an exact answer
- allow for typographical and misspelt words, etc.
- needless to say, it should be able to operate on several different domains without needing changes in the system structure, architecture or software.

Our approach is based on minimal parsing described in detail through case studies. This approach allows us to build a concept tree between word in the sentence this allows one to deal with nongrammatical sentences using syntacto-semantic categories and the keyconcept. We implemented several question answering systems using the minimal parsing keyconcept keyword concept. QA systems are an improvement over conventional search engines.

References

1. https://blog.kissmetrics.com/first-step-of-checkout/
2. Yi Chang, Hongbo Xu, Shuo Bai, A re-examination of IR techniques in QA system, in *IJCNLP 2004: Natural Language Processing—IJCNLP 2004*, pp. 71–80
3. Google, in http://www.google.com, website
4. C. Anantaram, Sunil Kumar Kopparapu, Chiragkumar Patel, Aditya Mittal: Repairing General-Purpose ASR Output to Improve Accuracy of Spoken Sentences in Specific Domains Using Artificial Development Approach. IJCAI 2016: 4234-4235
5. Maldovan et al., Trec8 (1999)

Chapter 6
Case Studies

As mentioned earlier, chatbots have become indispensable in the highly digitized and competitive world. More and more enterprises are taking advantage of being able to help their customers 24 × 7 and maintaining customer satisfaction and loyalty while still not draining their coffers to provide expensive human customer support. A sequence of question answering turns within a session could be loosely termed as a chatbot.

The core of minimal parsing systems is its ability to understand the question without formally parsing it by identifying keyconcepts and then identifying keywords to suit the semantics of the query. This minimal parsed representation is then used to retrieve the exact, or in the absence of exact information in its knowledge-base, answers that are semantically *close* to the query.

In the following sections, we briefly describe some QA systems (see the boxed text below) implemented by us, adopting the minimal parsing approach described in the previous chapters; we explore (a) their usability, (b) their language handling capability (lack of grammar as well as effectiveness across languages). Since these systems were implemented some years ago, we briefly describe the state of the art that existed at that time and the additional advantage that accrued using the minimal parsing approach.

© Springer Nature Singapore Pte Ltd. 2018

P. V. S. Rao and S. K. Kopparapu, *Friendly Interfaces Between Humans and Machines*, https://doi.org/10.1007/978-981-13-1750-7_6

A number of both textual and spoken natural language interfaces have been prototyped

the yellow pages interface allows queries such as

```
Where is a coffee shop in Thane?
```

the music search interface allows

```
I want to hear the song sung by Lata Mangeskar?
```

the Self-Help natural language interface for insurance allows queries such as

```
When will my policy with you mature?
```

or in a natural Hindi language interface (Mandi Bhav) a query like

```
Thane baazar mein aaloo ka kimat kya hai?
```

ठाणे बाजार में आलू की कीमत क्या है?

to name a few.

6.1 OASIS: Online Answering System with Intelligent Sentience

Sentience, according to the Webster's dictionary, is "the capacity for feeling or perceiving". Wikipedia [1] amplifies this further—as the capacity to feel, perceive or experience subjectively; according to the eighteenth-century philosophers; this concept was used to distinguish the ability to think (reason) from the ability to feel (sentience). Most human–machine interfaces (HMI's), such as question answering systems are built to "reason" the human query and respond with the most appropriate answer from their knowledge-base. However, the best user interfaces are those that are user friendly and go beyond merely being logical: meaning, while being logical, they "feel" for what the human is looking for and accordingly respond from their knowledge-base. In this chapter, we elaborate on how the "feel" can to a great extent be derived using the minimal parsing approach that we have described in detail in a previous chapter. This will be done mostly in the form of examples from several question answering systems that we developed, prototyped and demonstrated over a period of time at our organization as well as at conferences of peers.

Question answering (QA) systems are very important for information retrieval (IR) especially when humans are allowed to spontaneously query for information in their own (natural) language. Most QA systems function by analysing the query (text and, if speech, by converting it from speech into text using an automatic speech recognition (ASR) engine) to derive the "exact intent". Once the intent is deemed extracted, the QA systems try to "pattern" match the derived intent with one or more 'answers' from their knowledge-base. The intent extraction is mostly driven by the using the accepted grammar of the language while the "pattern match" is based on reason to compare the closeness of the intent pattern and the answer patterns. Consequently and more importantly, if no 'exact pattern' answers are available then the response is a simple "could not find what you were looking for".

However, a system that is based on "feel" would show graceful degradation when the intent pattern does not match any of the known answer patterns, namely, it should be able to provide the first best, the second best, the third best, etc. answers that are available in its knowledge-base and approximately match the query (intent) pattern especially in the absence of an exact answer pattern. As mentioned earlier, for example, "`Do you have Colgate toothpowder?`" could respond with "`No I'm sorry. I do not have that product.`" followed by "`I have Closeup toothpaste, will that do?`". This kind of performance by a QA system could be looked upon as being more user friendly, usable and intelligent. Clearly, for a QA system to function in this desired way not only requires knowledge of relationships between words (meanings) but also world knowledge (a taxonomy tree like a WordNet). All these aspects are realized in the minimal parsing framework that we have described earlier in this monograph.

We call this framework, *OASIS: Online Answering System with Intelligent Sentience*. The front end of this framework is a web interface with an option to input a query (see Fig. 6.1). The generic architecture that allows the QA system to demonstrate that OASIS has a feel for what the user wants is shown in Fig. 6.4. We will elaborate on the framework in detail.

'OASIS' is a question answering system developed at Cognitive Systems Research Laboratory (CSRL) of Tata Infotech in the late 90s. It uses innovative language processing techniques and answers questions asked in free English. It provides a way for the user to interact with the underlying system in natural English (free style—namely, natural, casual grammar in a transactional mode). It does not constrain the user to use a set of predefined keywords for accessing the system. This technology enables humans to use computers in much the same way that they would use a human assistant to get their work done, namely, without having to interact with computers in a predefined manner.

OASIS is aimed at evolving standard methods for creating question answering systems for different domains and with minimal user intervention. The main objective is to make a question understanding system general enough to work for any domain. This system retrieves from the knowledge-base the best answer that matches the question. As mentioned earlier, it provides the closest answer in the absence of an exact answer in the knowledge-base.

The main feature is the use of a WordNet kind of taxonomy tree, which has lexical relationships like synonyms, hypernyms, hyponyms, metonymy and homonyms. The other feature of the platform is minimal parsing, which uses with keyconcepts (or Cardinal Words) and Keywords (or arguments), to understand the intent of the query. Each keyconcept has a set of arguments, which are used to identify an answer. Keyconcepts are initially manually identified for a domain after analysing a set of frequently asked questions (FAQ's) and answers. Additionally, question type (or nature of the question) is also used in retrieving the answers. The knowledge-base has to be annotated manually or semi-automatically. The knowledge-base is tagged by headers in natural language rather than constructing specific keywords as headers.

A typical use case of OASIS is to enable quick access to information about an enterprise from its web pages (knowledge-base) as possible answers. The need for such a system can be understood based on the fact that an average user visiting any enterprise website has the following constraints:

- the user has little time, and does not want to be constrained by how he can or can not ask for information
- the user may not be grammatically correct all the time (he would tend to use transactional grammar)
- a first time user is unlikely to be aware of the organization of the web pages
- the user roughly knows what he wants and would like to query in the same manner as he would query another human in natural language.
- Additionally, the system should
  - be configurable to work with input in different languages
  - provide information that is close to that being sought in the absence of an exact answer
  - allow for typographical and misspelt words, etc.
  - needless to say, it should be able to operate on several different domains without needing changes in the system structure, architecture or software.

The front end of OASIS shown in Fig. 6.1 is a question box on the web page of an enterprise website. The user can type his question in natural English.

In response to the query, the system picks up specific paragraphs (present in the enterprise website and which are relevant to the query) and displays them to the user (see Fig. 6.2). Note that the output is an answer paragraph and the hyperlink option '1' and "2" show the links to the next best answers corresponding to the query.

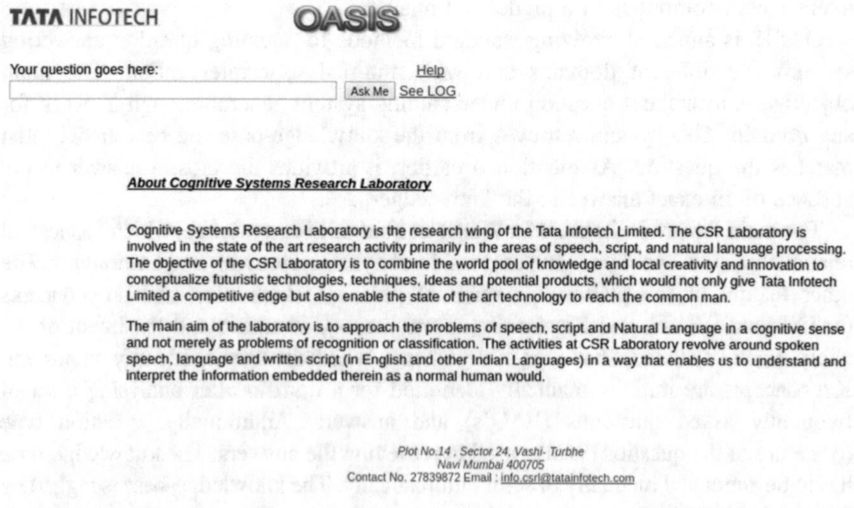

TATA INFOTECH OASIS

Your question goes here: Help

Ask Me See LOG

About Cognitive Systems Research Laboratory

Cognitive Systems Research Laboratory is the research wing of the Tata Infotech Limited. The CSR Laboratory is involved in the state of the art research activity primarily in the areas of speech, script, and natural language processing. The objective of the CSR Laboratory is to combine the world pool of knowledge and local creativity and innovation to conceptualize futuristic technologies, techniques, ideas and potential products, which would not only give Tata Infotech Limited a competitive edge but also enable the state of the art technology to reach the common man.

The main aim of the laboratory is to approach the problems of speech, script and Natural Language in a cognitive sense and not merely as problems of recognition or classification. The activities at CSR Laboratory revolve around spoken speech, language and written script (in English and other Indian Languages) in a way that enables us to understand and interpret the information embedded therein as a normal human would.

Plot No.14, Sector 24, Vashi- Turbhe
Navi Mumbai 400705
Contact No. 27839872 Email : info.csrl@tatainfotech.com

Fig. 6.1 OASIS front end

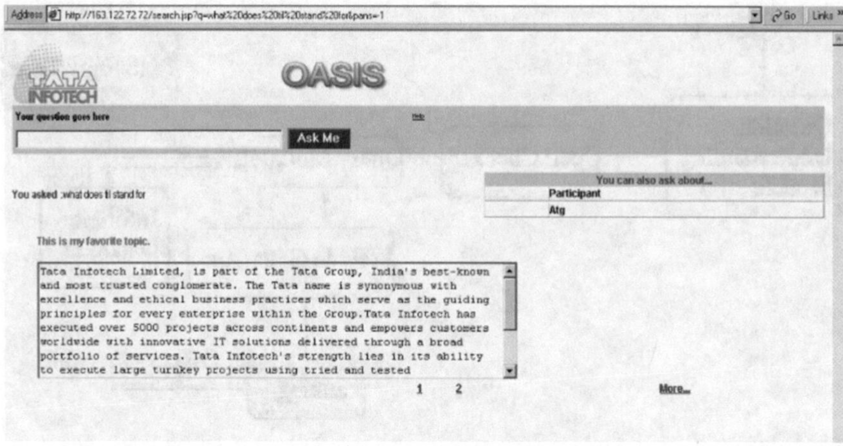

Fig. 6.2 Sample output. Showing the response to the query "`What does TIL stand for`"

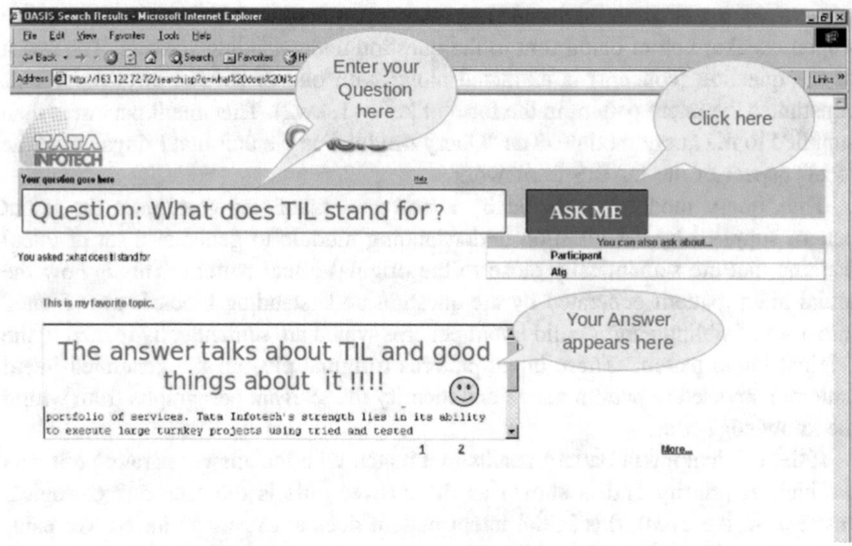

Fig. 6.3 Sample output in detail

Figure 6.3 shows the output in a little more detail indicating the location of the answer, the posed question, etc.

While providing an (exact or approximate) answer, the system also suggests related topics on which the user may ask questions if he likes.

As seen in Fig. 6.4, OASIS has what can be conventionally called a question understanding module driven by the minimal parsing approach; this allows the system to identify the intent of the query (determined by the extracted keyconcept

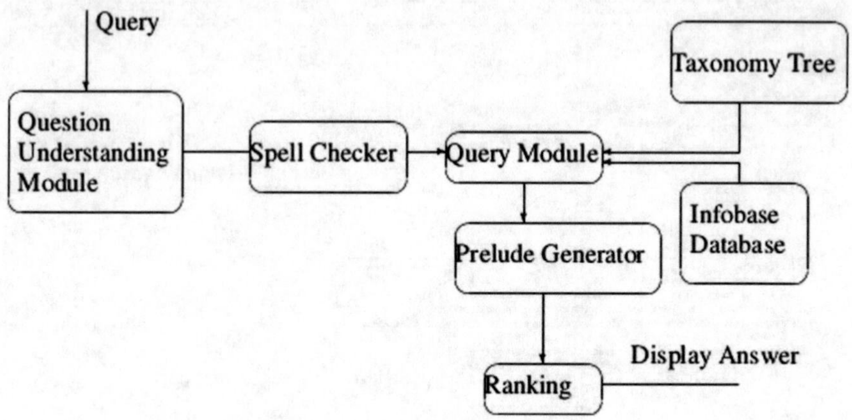

Fig. 6.4 OASIS architecture

and keywords in the query). The query is preprocessed (not shown in Fig. 6.4) to remove any punctuation marks and any known stop words. It is then passed through a spell checker before being sent to the question understanding module. The intent of the question (concept) is extracted along with one or several keywords. This constitutes the intent pattern in the form of kc(kw1, kw2). This intent pattern is then supplied to the query module. The "Query Module" is the unit that brings about the "feel" aspect of the OASIS framework.

The "query module", assisted by a powerful taxonomy tree, uses the intent pattern supplied by the question understanding module to generate a set of intent patterns that are semantically close to the original intent pattern. This is how the initial intent pattern generated by the question understanding block is now fanned into a set of sibling and/or child intent patterns which are semantically related to the original intent pattern. These intent patterns (original plus all the generated intent patterns) are used to pattern match and identify the relevant paragraphs from within the knowledge-base.

If the original intent pattern results in a match with the answer paragraph it gets the highest priority and is shown as the answer (this is the "reason" or logical answer). In the event, this initial intent pattern does not result in an answer paragraph, then the sibling and child intent patterns are used to "pattern match" to identify answer paragraphs. All such paragraphs of information picked up ("feel answers") are therefore appropriate to the initial query pattern and are then ranked in the decreasing order of semantic relevance to the query. The first ranked paragraph along with a (contextually appropriate) prelude generated by a prelude generator module is then displayed to the user. The preludes are of the form "You were looking for …, but I have found… for you" to indicate that the exact information that the user is seeking is unavailable.

Note that the "feel" answers are those that are triggered by the generated sibling and/or child intent patterns which fetch information most similar to the information sought by the user in the semantic sense.

For example a query like

"`Do you have Colgate toothpowder?`"

could respond with

"`No I'm sorry. I do not have that product.`"

followed by

"`I have Closeup toothpaste, will that do?`".

Note that the "feel" answer is because of the system's ability to see relationship between `toothpowder` and `toothpaste` and `Colgate` and `Closeup`.

In addition, OASIS maintains a session memory and can therefore connect questions within a session. This enables OASIS to 'complete' a query (in case the query is incomplete) using previous queries and their corresponding answers as reference. For example,

Question: `Who is Managing Director of Tata Infotech?`

MD	(Tata Infotech) =	?
Name	Name	Name
Type	Type	Type
Designation	Organization	Person

Question: `What does he do?`

? (he)

becomes

? (Managing Director of Tata Infotech)

Note that here "he" is resolved as "Managing Director of Tata Infotech" using its memory functionality and an answer referring to the work done by the managing director is given as response.

At the heart of the system is the taxonomy tree (see Table later in this section) and the answer paragraphs which are either ready at hand or are handcrafted for the particular domain that the QA systems needs to work for. As noted earlier, a taxonomy tree is essentially a structure which captures the relationship between different words. Typically, relationships like synonyms, hyponyms and meronymys are captured. The knowledge-base is the knowledge bank of the system and is manually engineered from the information available on the website of the enterprise. The knowledge-base essentially consists typically of a "paragraph of information" (which has a single central theme; as mentioned earlier a single keyconcept). These paragraphs are extracted from the information that exists on the web page. For example, the text appearing between <Answer> and </Answer> tags are paragraphs that are actually picked up from the website and the text between <option> </option> tag are the headers associated with the answer paragraph.

As mentioned earlier, these are manually engineered and sentences and not mere key-concepts and/or keywords. Note that there can be one or more headers for a paragraph.

<ANode>
<AnsNo>1601</AnsNo>

<header>
<option>Advantage of TIL</option>
<option>Definition of tata infotech</option>
</header>

<Answer>
Tata Infotech Limited, is part of the Tata Group, India's best-known and most trusted conglomerate. The Tata name is synonymous with excellence and ethical business practices which serve as the guiding principles for every enterprise within the Group. Tata Infotech's strength lies in its ability to execute large turnkey projects using tried and tested methodologies and established on-site and off-shore development models. Tata Infotech has operations in the areas of Software Services Outsourcing, System Integration Services, Contract Hardware Manufacturing and IT Education Services. The company provides end-to-end application lifecycle management services using its proven methodologies based on SEI CMM Level 5 and ISO 9001-2000 certified processes and has executed over 5000 projects worldwide. Tata Infotech specializes in seamless distributed on-site and offshore service processes, which deliver significant savings to the customers.
</Answer>
</ANode>

<ANode>
<AnsNo>1604</AnsNo>

<header>
<option>Achievements of TIL</option>
<option>Awards for TIL</option>
</header>

<Answer>
Tata Infotech's pioneering, often path-breaking, work has brought both rewards and recognition.
1. Top 10 Value Creators In India, by the Boston Consulting Group/Corporate Dossier
2. Voted No. 1 Systems Integrator in India by Voice and Data
3. World Economic Forum Award for being a pioneer in technology
4. ELCINA Quality Award for Manufacturing
5. IBM Excellence Award for ES CustomerView, our customer relationship management solution
6. CEU Award for Excellence In Exports
</Answer>
</ANode>

<Anode>
<AnsNo>1796</AnsNo>

<header>
<option>Availability of laptop permission template in the intranet</option>
<option>Procedure for laptop permission template</option>
<option>Link for laptop permission template in the intranet</option>
<option>Availability of laptop receipt acknowledgement template</option>
<option>Link for laptop receipt acknowledgement template in the intranet</option>

```
<option>Procedure for laptop receipt acknowledgement template</option>
<option>Laptop permission template</option>
<option>Laptop receipt acknowledgement template</option>
</header>

<Answer>
<option>Template for Permission to carry laptop is available in the following link:<br/>
</option>
<option><a href="http://163.122.32.xxx/NewIntranet/jsp/content/corporate/busiinfo/laptop
permission.doc">Laptop Permission</a></option>
<option>Template for Acknowledge for receipt of laptop is available in the following link:
<br/></option>
<option>    <ahref="http://163.122.32.xxx/NewIntranet/jsp/content/corporate/busiinfo/Laptop
Receipt_Details.doc">LaptopReceipt Details</a></option>
</Answer>

</Anode>
```


A sample taxonomy tree manually engineered is shown below. Some words are marked in green (KC) and pink (KW) to signify their occurrences in the examples later in this chapter.

Root	Synonym	Hyponym	Meronym
service_provider		isp	
training	training_programs	ntp,ctp,cbt	induction_programme, employee_ development,training_ schedule
employee_ development	employee_orientation	career_development,leadership_ development	employee_ development_ programs
career_development			
ctp	skill_upgradation		
sales	sell		supply,buy,purchase
logistics			procurement, distribute, delivery,orders,order_ forms
service	provide,solutions,support, servicing,facility	css,ites,outsourcing_services	manage
marketing	market_segments,market	overseas_marketing	sales
consultancy		software_consulting	
audit		finance_audit,quality_audit	inspection
inspection	scrutiny		
coe		bicoe,mscoe,baan_coe,hipaa_ coe,eai_coe,oracle_ coe,ebiz_coe,dss_coe	
database		db2,sql,informix,oracle,oracle_ applications,sas	
information_technology	software_company	oracle,compaq,sun,sequent,tandem,microsoft, nisc,baan,smartforce_systems,cisco,dell, human_factors,ibm,unisys,tata_burroughs, wausau,intel,novell,edexcel,cylink,hfi,sco, g_and_a_imaging,intershoppe,iona, business_objects, mapinfo,autodesk,northwood_geoscience, cardiff,bea_systems,software.com, lotus,captiva,j_and_b,	

(continued)

(continued)

Root	Synonym	Hyponym	Meronym
		ieoml,covad_communications,dos,catalyst_corporation,catalyst,tcs,solution_net,infoimage,pricewaterhousecoopers	
company		oracle,compaq,sun,sequent,tandem,microsoft,nisc,baan,smartforce_systems,cisco,dell,human_factors,ibm,unisys,tata_burroughs,wausau,intel,novell,edexcel,cylink,hfi,sco,g_and_a_imaging,intershoppe,iona,business_objects,mapinfo,autodesk,northwood_geoscience,cardiff,bea_systems,software.com,lotus,captiva,j_and_b,ieoml,rbi,bpcl,hpcl,cgwb,covad_communications,dor,dos,inarco,indo_asian_group,insma,krcl,RITES,lic,lufthansa,lufthansa_cargo,occ,soa,syndicate_bank,mfc,upsc,bhel,csd,catalyst_corporation,catalyst,ford_motor_company,northwest_airlines,united_airlines,sas,qantas,airindia,afc,big_w,mdor,consumer_electronics_group,tcs,yazaki,solution_net,govind_rubber_limited,banswara_syntex_limited,hth,givandev,worldbook,fedders_lioyd_corporation,rasbehari_enterprises,sulzer,indo_asian,sahara_india,sahara,tayf,great_eastern_shipping_limited,indian_aluminum_company,kewes,new_england_financials,metLife_insurance,kellogg,cnbc,ge_power_systems,orion_pharma,wausau_financial_systems,unichem_labs,great_eastern_shipping_limited,indian_aluminum_company,infoimage,ncc,tata_international,air_canada,nippon_airways,swissair,alitalia,indian_navy,insma,sitel,helmut_of_taj_exotica,stolt_nielsen,pricewaterhousecoopers,tata_yellow_pages,tata_infomedia,telecom_company	
activity	do,work,execute,involve,functions,business	marketing,research,training,project,service,audit,consultancy,administration,communication,education,manufacturing,operate,construction	
research		r_and_d,market_research,technical_research	research_projects
project		turnkey_projects,migration_projects,web_projects,systems_integration,functional_cost_management_system,y2k_project,porting_projects,maintenance_projects,study_projects,testing_projects,pilot_projects,research_projects,t_and_m,imaging_projects,compiler_projects,overseas_project	development,design,documentation,testing,installation,psc,reusable_component,module

These answer paragraphs and the taxonomy are important and enable the minimal parsing system to perform intelligently. Note that the words marked in green and pink are the ones that are used in the same queries that are analysed below.

The following are some examples from a fully functional system that was deployed for use within Tata Infotech for which this system was built. Notice that each question pattern results in answer header which are ranked in the decreasing order of semantic similarity with the question pattern.

Query: `oracle projects in til`

For this query, a part of the taxonomy kicks into place and this is used to form patterns which are close to intent pattern

```
projects(oracle)
```

Answer headers which are exactly or approximately close to projects(oracle) are chosen in decreasing order of semanticity, namely,

1. Project in oracle => projects(oracle)
2. Project for canteen in oracle => projects(oracle, canteen)
3. Project for DOR Montana in Tax solutions using Oracle => projects(oracle, tax solutions)
4. Project for TAYF in baan in Saudi Arabia => projects(baan, TAYF, Saudi Arabia)
5. Project for Unichem Labs in baan in India => projects(baan, Unichem, India)

In the above example, the first answer ("Project in oracle") is the perfect fit to the question. A non-minimal parsing system would stop as soon as this answer was found. However, in the absence of this answer, namely no answer header projects(oracle) associated with the system in the knowledge-base, the non-minimal parsing system would have returned an answer as "No answer found".

However, in the minimal parsing system, if that answer were not to be present then the next best answer would talk about "Project for canteen in oracle" or " Project for DOR Montana in Tax solutions using Oracle" which are specifically attached to the "Project in oracle" namely project done for "canteen" and for "DOR Montana", respectively. While both of them are indeed projects in Oracle, they have an extra snippet of specificity information which pushes their semantic closeness a little lower than "Project in oracle". If one looks at the last few answers, it is about a BaaN (similar to Oracle, see the taxonomy) project. This demonstrates the "feel" answers especially when the actual answers are not part of the answer knowledge-base.

Now let us look at another example, say a person wanting to join an enterprise ask the following query:

i am a ph.d. how can i join tata infotech

The answer headers chosen in decreasing order of semanticity using the taxonomy in the minimal parsing system are

1. Joining TIL for ph_d => join(ph_d, TIL)
2. Joining TIL with electrical engineering, electronics engineering or computer science engineering
3. Recruitment policy of TIL
4. Joining TIL by Referring Friend
5. Eligibility for joining atg
6. Link for recruitment

Thanks to the association between "ph_d" and " electrical engineering, electronics engineering, computer science engineering" the semantic similarity of the pattern join(ph_d)

is diluted to accommodate

join(electrical engineering)or join(electronics engineering) or join(computer science engineering)

with the last answer header "Link for recruitment"" which gives all the general information through a hyperlink to recruitment. The taxonomy that comes to play is shown below.

Root	Syn	Type
join	register,get_into,induction,enroll,enter,inscribe	
campus_interview	campus_recruitment	
selection	choose,opt_for,opt,prefer	
eligibility	suitability,criteria,factor,who_can,prerequisite	qualification,experience,skillset
qualification	educational_background,degree	graduate,post_graduate
post_graduate		mtech,mba,mca,ph_d,m_e
graduate		modern_engineering,classical_engineering,applied_arts,b_e,btech,masters
modern_engineering		computer_science_engineering,electrical_engineering,electronics_engineering
classical_engineering		civil_engineering,mechanical_engineering,aeronautical_engineering
civil_engineering	civil	
mechanical_engineering	mechanical	
aeronautical_engineering	aeronautical	
electronics_engineering	electronics	
electrical_engineering	electrical	

The use of taxonomy tree allows the QA system using minimal parsing system to output answers that are structurally and semantically close to the original intent of the query. We will explain this in greater detail in this example.

Let a user's original question be

"Can you let me know the activity in csrl"

Note that "csrl" is a word entity that is understood by the system thanks to the taxonomy tree.

Initially, the question ("Can you let me know the activity in csrl") is analysed to generate the intent pattern in the form of kc(kw, kw)

Post the removal of all the stop words, in this case, "Can", "you", "let", "me", "know", "the", "in" the question reduces to "activity csrl" which is

Original word	Word type
activity	kc
csrl	kw

So the original query is reduced to the intent pattern in the form of kc(kw,……), namely,

activity(csrl)

Now the system, by design, tries to find keywords, in this case, it searches for the parent of the word "csrl". The actual system log is shown below.

```
System | Step 1. SWITCHING KWS
System | Searching for PARENT of csrl...
System | Found : [atg]
System | New pattern Added*atg(kw)
```

The searching for the parent happens using the taxonomy tree which in this case finds the word token "atg" as the parent of the word "csrl". Using the newly found parent of the word csrl, namely atg, a new template is generated, namely,

activity(atg)

Now the QA system searches for the hyponyms of the word "csrl" in the taxonomy, followed by the siblings of hyponyms (namely, co-hyponyms) as well as the meronyms (the meronyms of the word "csrl" in the taxonomy) and co-meronyms as shown in the log of the system below.

```
System | Searching for TYPE of csrl ...
System | Searching for TO Siblings of csrl...
System | Searching for PART of csrl...
System | Searching for PO Siblings of csrl...
System | New pattern Added*isg(kw) [lpg#kw, network_security_group#kw, tirg#kw, iin#kw,
gis_group#kw, reengineering_group#kw, wireless_technology_group#kw]
```

As seen in the log, the system finds

[isg, lpg, network_security_group, tirg, iin, gis_group, reengineering_group, wireless_technology_group]

as the possible words (hyponym + meronym of csrl) from the taxonomy.

```
1>activity(kc)
2>isg(kw) Equiv-[lpg#kw, network_security_group#kw, tirg#kw, iin#kw, gis_group#kw,
reengineering_group#kw, wireless_technology_group#kw]
```

Using these identified words, a new template is generated, namely,

```
activity(isg)
activity(lpg)
activity(network_security_group)
activity(training)
activity(iin)
activity(gis_group)
activity(reengineering_group)
activity(wireless_technology_group)
```

The system now also searches by switching keyconcepts (kc) as shown in the log below.

```
System | Step 2. Switching Keyconcepts
System | Searching for Parent of activity...
System | Searching for Type of activity...
System | Found : [marketing, research, training, project, service, audit,
consultancy, administration, communication, education,
manufacturing, operate, construction, market_segments, market,
training_programs, servicing, facility, provide, solutions, support,
functioning]
```

Having found semantically equivalent keyconcepts, in this case,

`[marketing, research, training, project, service, audit, consultancy, administration, communication, education, manufacturing, operate, construction, market_segments, market, training_programs, servicing, facility, provide, solutions, support, functioning]`

The minimal parsing system then uses these keyconcepts to form patterns, namely,

```
System| New pattern Added*marketing(kc) [research#kc, training#kc, project#kc, service#kc,
audit#kc, consultancy#kc, administration#kw, communication#kc, education#kw, manufacturing#kc,
operate#kw, construction#kw, market_segments#kc, market#kc, training_programs#kc, servicing#kc,
facility#kc, provide#qt, solutions#kc, support#kc, functioning#kw]
1>csrl(kw)
2>marketing(kc) Equiv-[research#kc, training#kc, project#kc, service#kc, audit#kc, consultancy#kc,
administration#kw, communication#kc, education#kw, manufacturing#kc, operate#kw,
construction#kw, market_segments#kc, market#kc, training_programs#kc, servicing#kc, facility#kc,
provide#qt, solutions#kc, support#kc, functioning#kw]
```

```
marketing(csrl),
 research(csrl),
 training(csrl),
      ....,
functioning (csrl)
```

This is followed by looking for hyponym and co-hyponym of the keyconcept "activity" followed by the meronym of "activity, as shown in the log below

```
System | Searching for "Part of" activity...
System | Searching for "Type of siblings" of activity...
System | Searching for "Part of Siblings" of activity...
```

Using all this, the final set of templates that are constructed are

Template	Keyconcept	Keywords
0	activity	csrl
1	activity	atg
2	activity	isg Equiv-[lpg, network_security_group, tirg, iin, gis_group, reengineering_group, wireless_technology_group]
3	Marketing, research, training, project, service, audit, consultancy, administration, communication, education, manufacturing, operate, construction, market_segments, market, training_programs, servicing, facility, provide, solutions, support, functioning	csrl

These templates are then used, in this order, to fetch the corresponding answers, which is shown below.

Template	Answers Headers
0	1. 149—Activities of CSRL 2. 150—Activities of CSRL in Speech Synthesis 3. 151—Activities of CSRL in NLP 4. 1394—Activities of csrl
1	1. 1—Activities of atg 2. 125—Activity of atg with institute in research 3. 1394—Activities of atg in research 4. 1395—Activities of atg in Text Mining 5. 1396—Activities of atg in nlp for information retrieval 6. 1397—Activities of atg in Geometrics 7. 1398—Activities of atg in Embedded Systems 8. 1399—Activity of atg in Information Systems Security 9. 1400—Activities of atg in power 10. 1403—Activities of atg in Multimedia 11. 1404—Activities of atg in Reengineering 12. 1405—Activities of atg in operations research 13. 1406—Activities of atg in constraint programming 14. 1407—Activities of atg in Script recognition 15. 1408—Activities of atg in Bluetooth
2	1. 125—Activities of tirg 2. 143—Activities of iin 3. 1344—Activities of Reengineering group 4. 1408—Activities of wireless technology group 5. 1422—Activities of Internet security group
3	1. 149—Research in CSRL 2. 150--Research of CSRL in Speech Synthesis 3. 151—Research of CSRL in NLP

Now the answers are presented to the user of the query, namely, `activity (csrl)`, first from Template 0, then from Template 1 and so on. There are 27 answer headers that become eligible to be able to answer the query "`Can you let me know the activity in csrl`", namely,

1. 149–Activities of CSRL
2. 150–Activities of CSRL in Speech Synthesis
3. 151–Activities of CSRL in NLP
4. 1394–Activities of csrl
5. 1–Activities of atg
6. 125–Activity of atg with institute in research
7. 1394–Activities of atg in research
8. 1395–Activities of atg in Text Mining
9. 1396–Activities of atg in nlp for information retrieval
10. 1397–Activities of atg in Geomatics
11. 1398–Activities of atg in Embedded Systems
12. 1399–Activity of atg in Information Systems Security
13. 1400–Activities of atg in power
14. 1403–Activities of atg in Multimedia
15. 1404–Activities of atg in Reengineering
16. 1405–Activities of atg in operations research
17. 1406–Activities of atg in constraint programming
18. 1407–Activities of atg in Script recognition
19. 1408–Activities of atg in Bluetooth
20. 125–Activities of tirg
21. 143–Activities of iin
22. 1344–Activities of Reengineering group
23. 1408–Activities of wireless technology group
24. 1422–Activities of Internet security group
25. 149–Research in CSRL
26. 150—Research of CSRL in Speech Synthesis
27. 151– research of CSRL in NLP

Note that the answers headers while holding on to the semantics of the query asked in the Template 0 and are diluted as we go down to Template 1, 2, 3, etc. For example, the answer headers in Template 0 are in sync with what the user has asked, but later on, they start to dilute while holding on to the semantic of the query namely "`activities of csrl`".

This ability to stick to the semantic sense of the query and at the same time spread the net wide to give close but not exact answers to the query is possible because of the minimal parsing system duly helped by the taxonomy.

Note also that given the answer template one can search for statements in the database which conform to that template. This also indicates how one can form an SQL database query for a database. The advantage of this minimal parsing approach is that one can make sure that the answer provided is of the type expected.

6.2 QA System for a E-Book

We picked “The Fitness Factor: Every Woman’s Key to a Lifetime of Health and Well-being” a 314 pages book on fitness factor authored by Lisa Callahan who is the co-founder and medical director of the Women’s Sports Medicine Center at New York’s Hospital for Special Surgery to demonstrate the usefulness of building a QA system. A sample set of queries that we anticipated to be answered are given below, along with the answers provided by OASIS. From this, it is clear that the system is capable of dealing with existing sources of knowledge, with minimal preparatory work on the part of the designer. Note the preludes which are given out by the system; these are lively and relevant.

We show a list of queries that were asked during a pilot study and screenshots of the answers output by the minimal parsing system. Note that there is a taxonomy driving the system that is handcrafted.

For the query “`what are the benefits of exercise`” the system actually output the paragraph which describes about the advantages of exercises as its first option (see Fig. 6.5) while the second, semantically close, though not exact is related to “benefits of playing a new sport” (see Fig. 6.6). The essential “feel” answer is because of the system’s ability to see the relationship between sport and exercise in its taxonomy tree. Shown below are the answers that are semantically ranked answer headers.

- Benefits of exercise
- Benefits of playing a new sport
- Benefits of exercise after delivery
- Benefits of exercise for arthritis

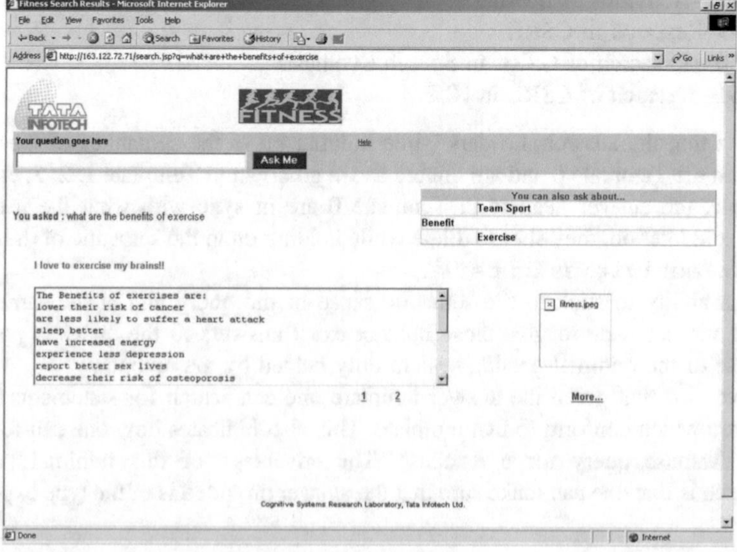

Fig. 6.5 Response to benefits of exercise

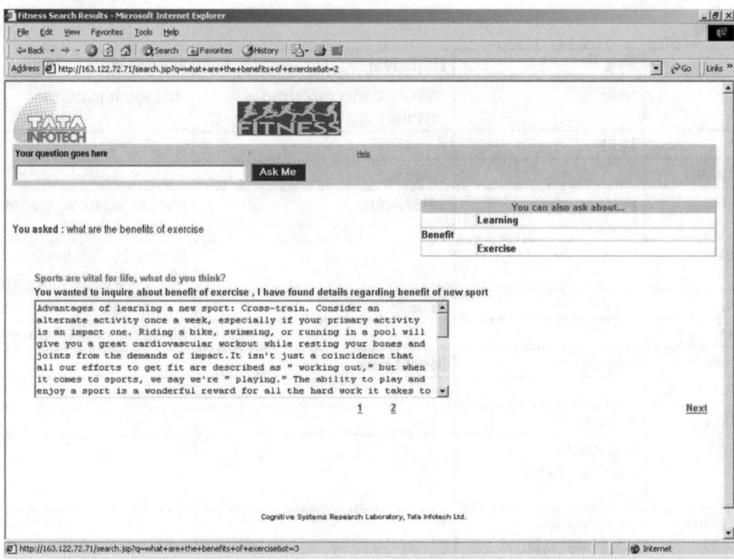

Fig. 6.6 Next best response to benefits of exercise

And a sample taxonomy which enables one to get the "feel" answers is shown below.

Root	Syn	Hyponym	Meronym
exercise(kc)	workout	diet,cardiovascular_exercise, yoga,sports,strength_training, weight_training,overtraining	tired,exercise_boredom
diet(kc)			weight_loss
sports(kc)		racket,bicycle,stair_stepper, treadmill	
bicycle(kw)		recumbent_bike,standard_bike	
health(kc)		mental_health,physical_health	energy,fitness
fitness(kc)		cardiovascular_fitness, strength_fitness	
disease(kc)		diabetes,arthritis, eating_disorders,heart_disease	
heart_disease (kc)			hbl,lbl
eating_disorders (kc)		anorexia_nervosa, bulimia_nervosa	
arthritis(kc)		osteoarthritis	
sports_injury (kc)		muscle_cramps,inflammation, stress_fractures, shin_splints,tendinitis	
treatment(kc)	cure		medicine
medicine(kc)		cortisone,	supplement,glycemic_index
drink(kc)		water	

(continued)

(continued)

Root	Syn	Hyponym	Meronym
food(kc)	fuel	carbohydrates,proteins,fat, vitamins,minerals,energy_bars	iron,calcium,calorie
fat(kw)	lipids		
vitamins(kw)		creatine	
human(kc)	person	woman,man	human_immune_system,heart, bone,muscle,
woman(kw)			pregrancy,delivery, after_delivery,child_care,child
child(kw)		infant	
heart(kw)			blood_pressure,cholesterol
reason(kc)	purpose	benefit,why	
why(qt)			
benefit(kc)	welfare		
factor(kc)	cause,risk, warning_signals, signs		
definition(qt)	what,describe, tell_about,brief, details, information,data, list,name, tell_me,give_me		

In response to the user query "`I have no time. How can I exercise`", the system exactly picks up a paragraph that is a response to this query (see Fig. 6.7) with a catchy prelude like "Time flies, isn't it". And for the question "`How can I exercise at home`", the response is shown in Fig. 6.7. And the response to "`How can exercise help during pregnancy`" is shown in Fig. 6.8. Similarly, for a query "`what are obstacles for exercising`", the system picks up answer paragraphs that are close in semantics to the query. In this case, the answer headers are

- 9 (Money is the problem for not exercising; Lack of money as a reason for not exercising; low-cost fitness exercise; how can I be fit without spending money),
- 5 (Lack of time as a reason for not exercising; Obstacles for exercising),
- 6 (Obstacles for not exercising)

Where the answers are

<ANode>
<AnsNo>9</AnsNo>
<Answer>
This is never an excuse not to exercise. Getting fitter and healthier doesn't have to cost you a cent. Consider checking exercise videotapes out of the library, or renting them from a video store. Begin a walking club in your neighbourhood or during your lunch hour at work. Look for reduced-rate memberships to gyms: Many offer discounts for off-peak hours or for family or group memberships.

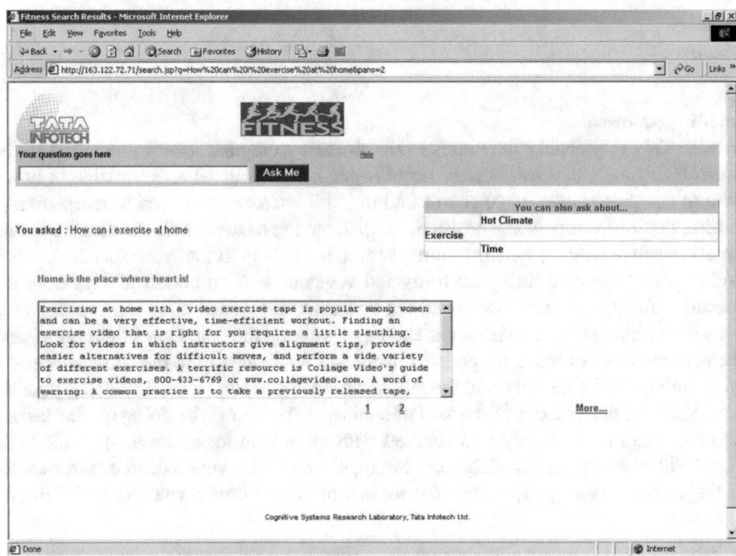

Fig. 6.7 Response to how can I exercise at home

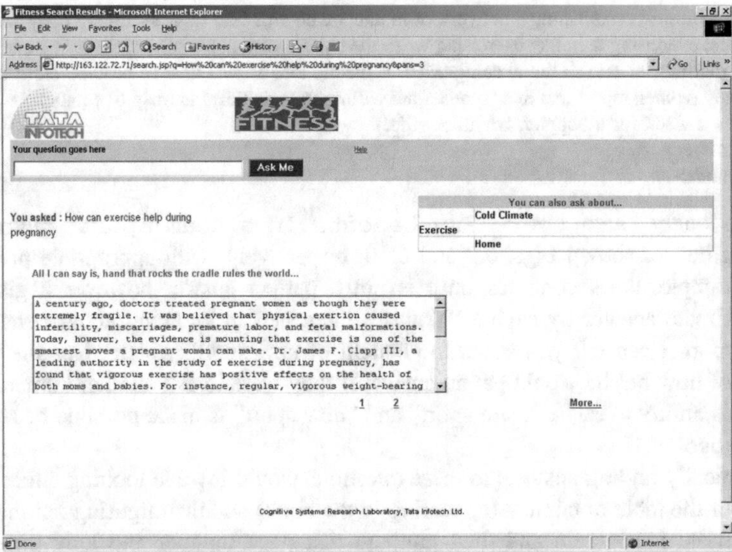

Fig. 6.8 Response to how can exercise help during pregnancy

Scan the classifieds in the newspaper for good-quality, little-used exercise equipment. Design your exercise program around inexpensive tools like a jump rope, or exercises such as push-ups and sit-ups that don't require any equipment at all.

</Answer>
</Anode>

<ANode>
<AnsNo> 5</AnsNo>
<Answer>This is probably the number one obstacle to regular exercise reported by busy women. Everyone's day has exactly twenty-four hours in it; what is variable is how we choose to use those hours. If you can't find time for exercise, you need to reorganize your priorities, putting your personal health right up at the top of the list. Since exercise is one of the most effective ways to pursue optimal health, it needs to fit into your schedule, just like brushing your teeth or eating your fruits and vegetables. If you don't always have thirty consecutive minutes to exercise, spend ten consecutive minutes three times a day in vigorous activity like taking the stairs, gardening, or walking the dog. Don't fight for a parking space near the front of the mall: park far away and walk quickly to the entrance. Mow the lawn in summer; rake the leaves in the fall. Walk briskly around the house doing your daily chores. See how fast you can fold a load of laundry. Who knows, by doing routine tasks at a brisk pace, you may suddenly find yourself with more time for exercise, after all! Make a contract with yourself; or schedule exercise appointments on your calendar, and treat them as if they were doctor appointments (or some other commitment you wouldn't ignore).
</Answer>
</Anode>

<ANode>
<AnsNo>6</AnsNo>
<Answer>This is a commonly identified barrier to a busy woman's commitment to a regular exercise program. Some solutions to this barrier include: finding a gym with child care facilities; establishing a network of mothers who will take turns swapping baby-sitting duties; exercising at home; exercising when the children are in school, or exercising early in the morning or late in the evening when someone else is available to provide child care. (Many women report that their spouses are willing to trade thirty to forty-five minutes a few times a week for a happier, healthier wife!)
</Answer>
</Anode>

The query "`Are there any benefits in a team sport`" returns the answer header shown Figs. 6.9 and 6.10, respectively, with appropriate preludes. For example, the second response is not a perfect answer however it gives an approximate answer by saying "You are interested in knowing about benefits of team sport. I can tell you about the benefit of new sport" which gives the user a sense of how he/she would get an answer if they were to ask another human. Note that this ability to relate "team sport" and "new sport" is made possible because of the taxonomy.

Typically finding answers to these questions would involve looking either at the index or the table of content (appearing in the book) and then figuring out the most appropriate index item and then jump to that page number and look through a couple of pages to get to the answer. This is definitely a laborious process. The QA system on the other hand should be able to make it easy and usable.

A few more minimal parsing (OASIS) system responses to actual queries are given in Appendix 9.2 Sample HMI (e-book on Fitness).

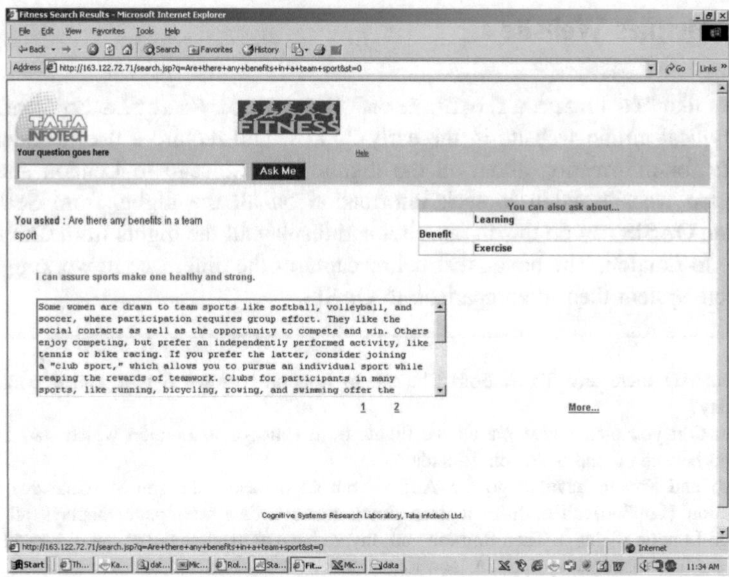

Fig. 6.9 Response 1 (Are there benefits of team sport)

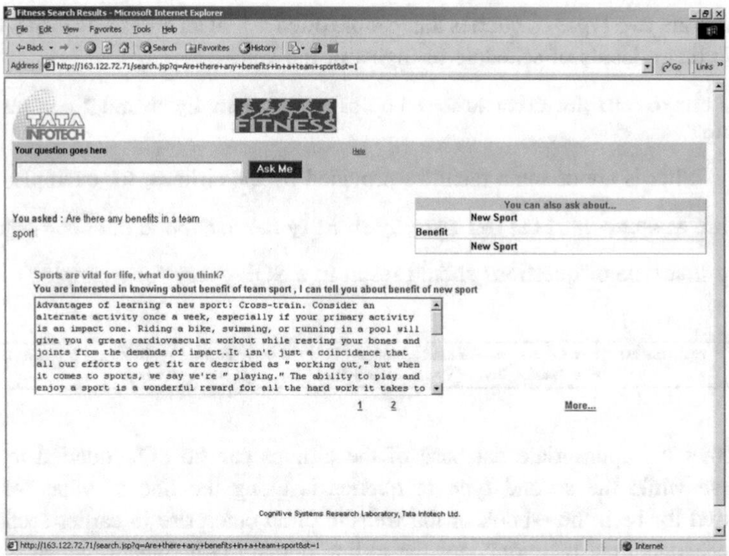

Fig. 6.10 Response 2 (Are there benefits of team sport)

6.3 Airlines Website

Queries like "`Is there a flight from Chicago or Seattle to London?`" on a typical airline website in the early 2000 would require a user to query the website for information about all the flights from Chicago to London and then query the website again to seek information on all the flights from Seattle to London. OASIS can do this in one shot and display all the flights from Chicago or Seattle to London. The boxed text below captures the difference in working of the available system then in comparison to OASIS

> Ramesh: Are there any flights from Chicago to London between 6 p.m. and 9 p.m. on Thursday?
> Shyam: Can you please give me all the flights from Chicago to London which start from Chicago between 6 and 9 PM on Thursday?
> Ramesh and Shyam have to go the Airline website ... select the source (Chicago) and destination (London) cities from a drop down menu ... the web page displays all the Chicago-London flights ... Then Ramesh and Shyam have to manually scan the displayed list to get the information they seek. A laborious process.
> OASIS: Displays only the information of flights from Chicago to London which are functional on a Thursday between 6 and 9 PM.

There are two types of queries that people often ask of an airlines website. One is to give them details of schedule of flights of the kind

"`Is there a flight from Mumbai to Delhi between 4 p.m and 7 p.m on 25 June?`"

and the other is about some facilities provided by the airlines, for example,

"`What are the facilities for physically handicapped passenger`".

The first type of questions should result in a SQL query for example,

```
Select * From Database Where Origin = 'mumbai' and Destination = 'delhi' and Departure_Time >= #16:00# and Departure_Time <=
#19:00# and ( Days_of_Operation like '%5%' or Days_of_Operation = 'Daily' )
```

so that the appropriate database of the airlines can be SQL queried to get a response while the second type of queries is along the line of what we have discussed for be it the e-book or the website of an enterprise in earlier sections.

The handcrafted taxonomy for the airline domain (Air Sahara, which is since defunct!) is shown next.

Root	Class	Synonym	Hyponym	Meronym
service	kc		value_added_service,cargo_service, in-flight_service,helicopter_service, trolley_service	
facility	kc	amenity, provision, advantage, benefit	business_lounge,tele_check_in, reading,wings_wheels,valet, pick_up	
cargo	kw	consignment		
airsahara	kw	sahara		
safety	kc			precaution,caution, protection
handle	kc			
weight	kc			
payment	kc		credit_card,cash	
procedure	kc	steps		rule
limitation	kc	restriction	prohibit	
goods	kw	thing,material, article,item	chemical	
valuable	kw	invaluable		
loss	kc			
animal	kw		pet,cat,dog,rabbit	
life_and_death	kc		live,dead	coffin
live	kc	alive		
dead	kc	no_more		human_remain
rule	kc	condition, prerequisite, regulation		
cost	kc	fees,fare,charge		payment
carry	kc	takeaccompany		
in_flight	kw	on_board, onboard, on_flight, in-flight		
shipment	kw			
wet	kw	liquid		
customer	kw	passenger,client		
care	kc			
satisfaction	kc			
reading	kw		magazine,newspaper,novel	
recreation	kw	entertainment	reading	
passenger	kw		tourist	
staff	kw			executive,pilot, air_hostess

Some of the queries that were pilot tested.
Example of SQL type queries picked from the log generated by the QA system.

`Is there a flight from Mumbai to Delhi between 2 p.m and 4 p.m on Wednesday? Is there a flight from Mumbai to Delhi between 2 p.m and 4 p.m on 12th january 2005?`

```
Select * From Database Where Origin = 'mumbai' and Destination = 'delhi' and Departure_Time >= #14:00# and Departure_Time <= #16:00# and ( Days_of_Operation like '%3%' or Days_of_Operation = 'Daily' ) and Origin = 'mumbai' and Destination = 'delhi' and Departure_Time >= #14:00# and Departure_Time <= #16:00#
```

`Can you give me information about flights from Mumbai to Delhi between 2:00 p.m. and 6 p.m.?`

```
Select * From NewMainSchedule Where Origin = 'mumbai' and Destination = 'delhi' and Departure_Time >= #14:00# and Departure_Time <= #18:00#
```

`Is there a flight from mumbai to delhi which reaches delhi before 7 p.m. or after 11p.m.`

```
Select * From NewMainSchedule Where Origin = 'mumbai' and Destination = 'delhi' and Destination = 'delhi' and ( Departure_Time <= #19:00# or Departure_Time >= #23:00# )
```

`Can you give me information about flights from mumbai to delhi between 6 pm and 9 pm?`

```
Select * From NewMainSchedule Where Origin = 'mumbai' and Destination = 'delhi' and Departure_Time >= #18:00# and Departure_Time <= #21:00#
```

`Is there any flight from New Delhi on Sunday which reaches between 4 p.m. to 6 p.m.?`

```
Select * From NewMainSchedule Where Origin = 'delhi' and ( Days_of_Operation like '%7%' or Days_of_Operation = 'Daily' ) and Arrival_Time >= #16:00# and Arrival_Time <= #18:00#
```

`Is there a flight from Mumbai to Delhi between 2 p.m and 4 p.m on 12 january 2005?`

```
Select * From AS_IA_JET_Sample Where Origin = 'mumbai' and Destination = 'delhi' and Departure_Time >= #14:00# and Departure_Time <= #16:00#
```

`Is there any flight from New Delhi on Sunday which reaches between 4 p.m. to 6 p.m.`

```
Select * From NewMainSchedule Where Origin = 'delhi' and ( Days_of_Operation like '%7%' or Days_of_Operation = 'Daily' ) and Departure_Time >= #16:00# and Departure_Time <= #18:00#
```

`is there a flight from mumbai to delhi which reaches before 7 p.m. or after 11p.m.`

```
Select * From NewMainSchedule Where Origin = 'mumbai' and Destination = 'delhi' and Arrival_Time <= #19:00# and
( Departure_Time <= #19:00# or Departure_Time >= #23:00# )
```

`Give me details of all the flights from Mumbai to Delhi between 9 am and 1700hrs`

```
Select * From NewMainSchedule Where Origin = 'mumbai' and Destination = 'delhi' and Departure_Time >= #9:00# and
Departure_Time <= #17:00#
```

`Is there any flight from Delhi to mumbai on Sunday between 4 p.m. to 6 p.m.`

```
Select * From NewMainSchedule Where Origin = 'delhi' and Destination = 'mumbai' and ( Days_of_Operation like '%7%' or
Days_of_Operation = 'Daily' ) and Departure_Time >= #16:00# and Departure_Time <= #18:00#
```

`Is there a flight from Mumbai to Delhi between 4 p.m and 7 p.m on 25/6/2004?`

```
Select * From AS_IA_JET_Sample Where Origin = 'mumbai' and Destination = 'delhi' and Departure_Time >= #16:00# and
Departure_Time <= #19:00# and ( Days_of_Operation like '%5%' or Days_of_Operation = 'Daily' )
```

`Is there any flight from Delhi to mumbai on Sunday between 4 p.m. to 6 p.m.`

```
Select * From NewMainSchedule Where Origin = 'delhi' and Destination = 'mumbai' and ( Days_of_Operation like '%7%' or
Days_of_Operation = 'Daily' ) and Departure_Time >= #16:00# and Departure_Time <= #18:00#
```

Sample informative questions responses are given below

Question: `facility for physically handicapped passenger`
Answers

#1 27 (Facility for passengers with restricted mobility, Facility for handicapped passengers, Facility for senior citizens. Facility for wheel chair, Help for handicapped, Help for old, Assistance for handicapped people, Special features for the handicapped)

#2 19 (Advantage of airsahara, Value-added Services of airsahara, OnGround Services of airsahara, Facilities provided to passengers. Facilities of airsahara)

#3 24 (City check-in, Facility to make check-in easy, Convenience for passengers travelling with hand baggage, Facility for passengers with hand baggage, procedure for check-in, Boarding pass for passengers with hand baggage)

<ANode>
<AnsNo>27</AnsNo>
<Answer>Passengers with restricted mobility are given extra assistance. In case of a senior citizen customer service staff escorts him. If a passenger requires a wheel chair he/she is provided with that facility as well.
</Answer>
</Anode>

<ANode>
<AnsNo>19</AnsNo>
<Answer>Air Sahara believes in delivering the best to all our customers. It is this belief that is reflected in all our services. Flying Air Sahara has a definite advantage because we believe in adding value to your travel.Air Sahara offers its customers a variety of value added services that include:-
VALET SERVICE
TELE CHECK-IN SERVICE
FASTEST BAGGAGE RETRIEVAL
ADVANCED SEAT RESERVATION
INTERACTIVE VOICE RESPONSE
AUTOMATED FLIGHT ARRIVAL/DEPARTURE INFORMATION
LOUNGE FACILITIES
CITY CHECK IN
24 HOUR RESERVATION
AIRPORT TRANSFERS
SPECIAL FEATURES FOR CHILDREN AND INFANTS
</Answer>
</Anode>

<ANode>
<AnsNo>24</AnsNo>
<Answer>We at Sahara believe in providing you with maximum comfort. Which is why we have introduced our new City Check-in facility for passengers travelling with hand baggage. Now you don't have to arrive hours before take-off. Just drop in our City Check in office at a convenient time one-day before departure and collect your boarding pass. The Sahara City Check in facility is currently available at Delhi, Mumbai, Bangalore, Chennai and Luknow.
</Answer>
</ANode>

Question: `What is the procedure for check in`
Answers

#1 18 (Tele Check-in information, Telephone numbers to check-in, Facility for check-in. Procedure for check-in, Check-in made easier in airsahara. Access facility of airsahara, Contact for airsahara)

#2 24 (City Check-in, Facility to make check-in easy, Convenience for passengers travelling with hand baggage, Facility for passengers with hand baggage, procedure for check-in, Boarding pass for passengers with hand baggage)

<ANode>
<AnsNo>18</AnsNo>
<Answer>Sahara is the first domestic airline to offer Tele-check in facility for both Economy andamp; business class passengers. We now have dedicated lines on which you can access information on flights, timings, Frequent Flyer Programme and even ask for a seat of your choice. All this is to ensure that you can access us at Sahara anytime, everytime and at your convenience. Delhi 011-23359801 Mumbai 022-26156363 Bangalore 080-5220626 /5220665 Chennai 044-2561643 /2561544 for further details please contact:- Delhi 011-23359801 Mumbai 022-22836000 Bangalore 080-5584457/3937/3897 Chennai 044-7110202 Kolkata 033-22826118-22 Lucknow 0522-2334426/4428 Patna 0612-2239569/2722/2723 Pune 020-6059003/9004/9005 Hyderabad 040-23223767/23212237 Goa 0832-2230237,2230634 Guwahati 0361-2548676,2547808 Varanasi 0542-2507872/873 Gorakhpur 0551-2204793/94/95 Allahabad 0532-2608533/34/35
</Answer>
</Anode>
<ANode>
<AnsNo>24</AnsNo>
<Answer>We at Sahara believe in providing you with maximum comfort. Which is why we have introduced our new City Check-in facility for passengers travelling with hand baggage. Now you don't have to arrive hours before take-off. Just drop in our City Check in office at a convenient time one-day before departure and collect your boarding pass. The Sahara City Check in facility is currently available at Delhi, Mumbai, Bangalore, Chennai and Lucknow.
</Answer>
</ANode>

Question: `how do you ensure a passengers safety?` Or
`Tell me about Safety of cargo in airsahara`
Answers:

#1 1 (Cargo services by airsahara, Safety of cargo in airsahara, Types of cargo handled by airsahara, Cargo carried by airsahara)
#2 16 (Safety in airsahara, Maintenance of aircraft, Skills of Engineering department)
#3 34 (Helicopter service of airsahara, Travelling to a remote destination, Facilities offered in Sahara helicopters, Safety measures in Sahara helicopters)

<ANode>
<AnsNo>1</AnsNo>
<Answer>A new high in the domestic cargo arena. Air Sahara meet all needs for transportation of cargo. We can take care of all transport requirements from normal cargo to perishable and even pet animals. Air Sahara will be introducing stickers for all its consignments to ensure the smooth handling and tracking of cargo. It is the first private airline in India to install X-ray Machine in Delhi Mumbai to screen the consignments to avoid opening the consignments physically, thus ensuring the shipments are delivered intact. Customer satisfaction of our Agent/Clients is the key to our success. Many new systems and swift movement of cargo was introduced by Air Sahara thus making a revolutionary change in the Cargo Industry and thus became the Trend Setter in many fields. Each and every consignment thoroughly travel through professional hands and maximum precautions

are taken to deliver the consignment safely to our clients. It is a matter of pride to announce that Air Sahara claims for mishandling/loss/damage/of consignments. This proves the service standards and the client satisfaction level which itself is a record.
</Answer>
</Anode>

<ANode>
<AnsNo>16</AnsNo>
<Answer>Safety is of prime importance at Sahara, and we are the only domestic network with sophisticated, imported equipment for thorough baggage and aircraft checks. With DGCA approved maintenance shops, NDT shops, overhauling and test facilities, we offer world class care to our customers. Our reliable service has won us the trust of an International airline like Lufthansa, which has entrusted the maintenance, and 3 `C' check inspection of their aircraft to us. Every year we invest over Rs. 50 crores on aircraft maintenance and computerized inventory to ensure dispatch reliability and on-time performance. We have a fleet comprising of Boeing 737 - 400 and four new, highly advanced Helicopters (Dauphins and Ecuriel). Our engineering department is constantly upgraded with highly advanced test equipment required for routine maintenances and heavy checks like `C' and `D' checks, engine changes etc. Engineering Department forms on integral part of Airlines. It ensures timely dispatch of a safe aircraft thereby ensuring a good dispatch reliability.
</Answer>
</Anode>

6.4 Natural Language HMI to Yellow Pages

Yellow Pages are a source of information about various commercial organizations: their addresses, phone contact and other details, including information regarding the commercial activity of the organization. These are very useful and frequently used by individuals as well as other business houses till the advent of the Internet and web pages as a popular medium for such needs. Early on, the only way to access information from these Yellow Pages directories was to physically look into a huge hard copy directory, which was not only laborious but also time consuming. It also required the user to be familiar with the organization of the physical (book) directory.

This was followed by interactive voice response (IVR) based contact centres that helped the users to query information orally by telephone, by asking an operator. While this was easier than browsing through the physical directory, it still has several pitfalls. The time spent on trying to get the information was still quite long, often involving waiting in a queue on telephone; at the end of the transaction, one was still unsure if one had in fact obtained the information that one was looking for.

Subsequently, enterprises that had access to Yellow Pages information in the digital form went a step further; they allowed people to search through their electronic directory over the web. While the baton was back in the hands of the user, there was still an element of skill required to search for what one was looking for.

In the early 2000s, we built an SMS-based natural language interface to query an electronic Yellow Pages directory [2] based on the concept of minimal parsing which we call YPSMS. In this section, we describe the Yellow Pages search enabled by sending a text message using the short message service (SMS). The central idea behind the system was to have a user friendly and usable interface, which did not impose any constraint on the way a user could query the Yellow Pages directory. Our implementation was for messages in natural English, but given the use of minimal parsing, this could work for any language. The system understands the intent of the query using minimal parsing and "intelligently" searches the Yellow Pages directory to retrieve information. This retrieved information is then sent back to the user in the form of a SMS.

6.4.1 History

As we look back today, mobile phones have made significant inroads into the society and there is a large population that is mobile. However, in the late 1990s and early 2000s when mobiles were relatively new, the competition between the mobile service providers, to retain their current subscribers and attract new subscribers was to provide them with value-added services (VAS) which was probably the only way, then, to increase the average revenue per user (ARPU). The mobile service providers were on the lookout for applications that were not only innovative but also useful in day to day life. In India, like most of the developing countries then, there was a trend to use SMS more than voice because of economic considerations.

Yellow Pages directory is a very informative resource that provides up-to-date information about commercial organizations. It is very common for a directory to be available for every town or city and it is very often the only authentic resource to get information about commercial enterprises. To start with the physical Yellow Pages directory was the only source of information. To get information, a user browsed through the directory and got to the information that he was looking for, mostly by relying on the index. Often, access to the information was dependent on the organization of the Yellow Pages directory and how current the information was among other things. These directories were printed at the most once a year which mean that information was not very up to date. Unless very familiar, a user would need to take effort and time to get to the information. Enterprises publishing Yellow Pages saw the need to update their database more frequently than they could print and distribute the hard copy directories. They were therefore motivated to enable a central place where the data could be updated more frequently. This led to the establishment of interactive voice response (IVR) systems, which were accessible to the user. This was more convenient for the user in the sense that he just needs to make a phone call and "ask" for information; a live agent would search (on behalf of the user) the Yellow Pages electronic database directory using a series of SQL queries and convey the information back on the phone to the user.

While it was a nice solution, it was

1. Time consuming (very often one has to be in the queue listening to advertisements or a very irritating "Please be on hold, your call is important to us and we will get back to you as soon as one of our agents is free to take your call")
2. Expensive for the user (telephone bill for the whole duration of the call (Including the time taken to get to the live agent!)
3. Highly dependent on the searching skill of the live agent. In addition, acoustic confusions could prevail because of

    1. the noisy and low bandwidth telephonic conversation and
    2. pronunciation and accent further prolonging the interactive session between the user and the live agent and

4. Expensive for the service provider (the need to set up a call centre and engage 24×7 live agents)

Once the first mobile phones entered the scene, especially in developing countries like India, SMS was the most used functionality on the mobile device, thanks to the exorbitant cost of the voice channel. It was in this scenario, for reasons mentioned earlier, that YPSMS (our SMS enabled searching the Yellow Pages directory using natural English) was conceptualized based on the minimal parsing described earlier in this monograph. While exploiting the minimal parsing concept, the system aimed at the following salient features which were significantly ahead of its time from the usability perspective.

- System should be easy to use
- Should not require the user to remember any specific code or mnemonic to query the directory
- In the absence of an exact match to the query (in the knowledge-base), the system should provide the user with the 'next best fit' answer in some sense
- Should cater to SMS lingo and typographic errors that might occur when writing an SMS using a feature phone (smart phones had not yet made their entry!)

6.4.2 System Overview

The system consists of essentially three main modules. The first module interfaces with the SMS gateway of the telecom operator and passes the SMS query as a text string to the second module, which is the heart of the system and is based on the minimal parsing procedure described earlier. This module analyses the natural English queries (including SMS lingo and automatically generates a set of kc-kw patterns). It then passes these patterns on to the next module which is the database query module. The output of the query module is one or more answer text strings, retrieved from Yellow Pages knowledge-base. These are then sent to the user as one or more SMS depending on the length of the string retrieved.

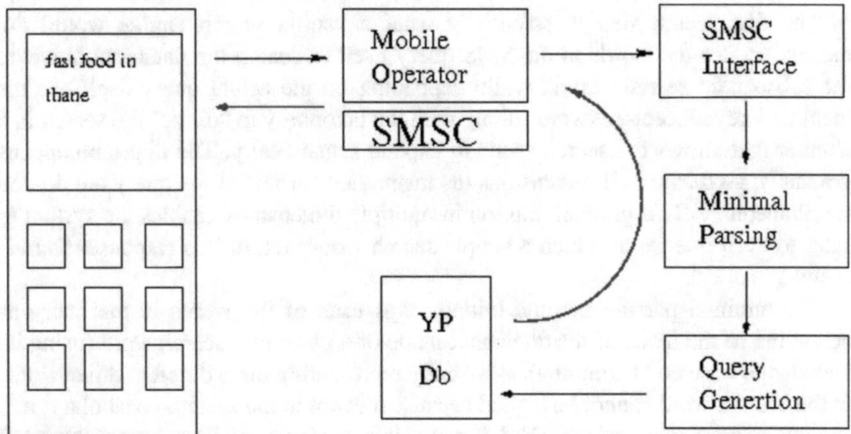

Fig. 6.11 YPSMS architecture

Figure 6.11 gives the overview of the system that is able to minimal parse the SMS query to understand the intent of the query and then intelligently access the Yellow Pages directory on the mobile network using SMS.

The SMSC module interfaces with the SMSC (SMS centre) of the mobile service provider. Its main functionality is to obtain the SMS query from the SMSC and then pass on the text query as is to the minimal parsing module, which is the heart of the system. The minimal parsing module understands the intent of the query using the keyconcept-keyword terminology rather than treating the SMS string as a set of keywords. Having identified the keyconcept and keywords, the query generating module generates a list of possible search criteria which can be used to generate an SQL query to extract Yellow Pages listings from the database.

It should be noted that, in the Yellow Pages query domain, the default keyconcept is 'find address' if it is not explicitly stated. For example, if the query is "Studio in Andheri" what one actually means is "[(find |tell me) address (of a)] Studio in Andheri". In this domain, we therefore need not specifically look for a keyconcept; we concentrate on the keywords, with the understanding that the keyconcept is implicit. It is also to be noted, as mentioned in earlier chapters, that people generally do not construct grammatically and semantically complete query statements in a transactional like scenario as querying Yellow Pages. This is even more applicable when composing an SMS on a mobile phone, as is the case in this Yellow Pages context. The dimensionality of the keywords is determined by the nature of the implicit keyconcept. In this case, these would be, for instance, the nature of the article (food, apparel, electronics, jewelery etc.) or organization (courier company, cinema house, hospital, mall, bank, etc.) and the location (Bandra, Dadar, North or new Mumbai, etc.). There could be other keywords such as working hours (24 by 7, office hours only), e-mail addresses, etc.

The first search step is essentially what a vanilla search engine would do, namely, to use the words in the SMS query itself to search the database; however the subsequent search criteria while depending on the actual query itself use the identified keyconcept-keywords along with the taxonomy to "dilute" the search in a manner that allows the search space to expand semantically. The dilution happens gradually, so that not all dimensions (as mentioned earlier) of the query are diluted simultaneously. This gradual dilution in multiple dimensions enables the system to cater to even queries to which a simple search would return "no responses found" answer.

The minimal parsing module initially tags each of the words in the query as belonging to the name of the company, name of a place or a search word (using its knowledge-base and taxonomy), essentially representing three different dimensions. In the event a word cannot be tagged because it is not in the system vocabulary; it is tagged as unknown and checked for possible spelling mistakes using the spell checker module; then an attempt is made to tag the corrected word appropriately. The tagged information (T1) is sent to the query module, which forms a SQL query to retrieve information from the database. In case no records are returned from the database, another (diluted) tagged list (T2) is generated using the domain taxonomy; this altered set of words is then used by the query module to search the database. This process is continued until one or more records are returned from the database to the query posed by the user. For example, let us say there is an enquiry regarding a particular firm (say DTDC) from a particular location, it first dilutes this to cover similar firms at that same location. If this does not work, it might dilute the location specification next and try for the same firm at other neighbouring or easy-to-reach locations. The generation of the tagged list (T1, T2, …) depends on the initial query; the aim is that the subsequent SQL queries have higher probabilities of attracting records in the database. This systematic and progressive dilution of the search allows the system to answer intelligently (as a normal human would do), by giving exact answers when present in the database or approximate but close and usable answers in the absence of an exact answer in the database.

6.4.3 Advantage of a User-Friendly Natural Language Interface

Designing a strategy for generating T1, T2, … from the initial query of words is crucial and made possible because of the concept of minimal parsing which is able to identify the dimensionality of the keywords. This assignment of semantic dimensions with the support of a domain specific taxonomy helps in digging out information from the database in the event of information being absent for the initial query patterns or words. This strategy gives the system an edge over using an ordinary keyword-based search strategy.

A minimal parsing system described earlier in detail needs to (a) identify key-concepts and keywords in a query and then (b) use a taxonomy tree to enable semantic expansion of the identified keyconcepts and keywords. While other aspects like being able to correct spelling mistakes are required, as such they are not central to the minimal parsing system.

An indicative (and very partial) taxonomy is shown in the map for the central word "food". Clearly, there are synonyms (example "sweet | mithai") and there are types of relationships (example "vermicelli" is a type of "noodle") at different levels. This taxonomy is particularly useful in the minimal parsing system to enable it to provide answers that are semantically close to a particular query. For example, if someone asked for burger, the taxonomy would allow us the knowledge that pizza is also like burger because both are types of fast food. This allows the system to provide usable and semantically relevant answers in the absence of exact responses in the answer domain (in this case the Yellow Pages database).

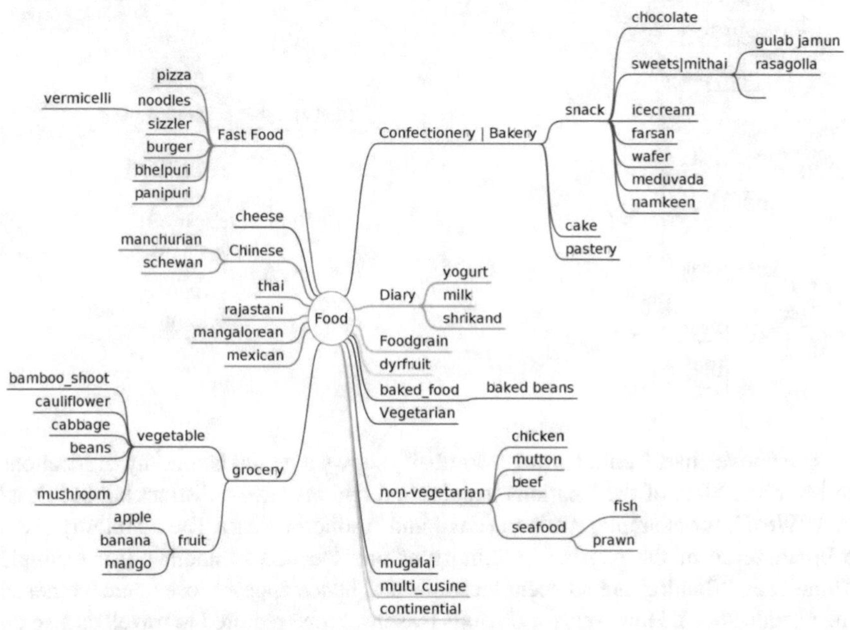

In the Yellow Pages domain, there is a need for a taxonomy for locations or areas of a city. For representational purposes, we used a distance matrix to enable identifying neighbourhoods of a given location. Note that one could easily use Google Maps API [3] to enable this functionality now though it was not in existance when we prototyped this.

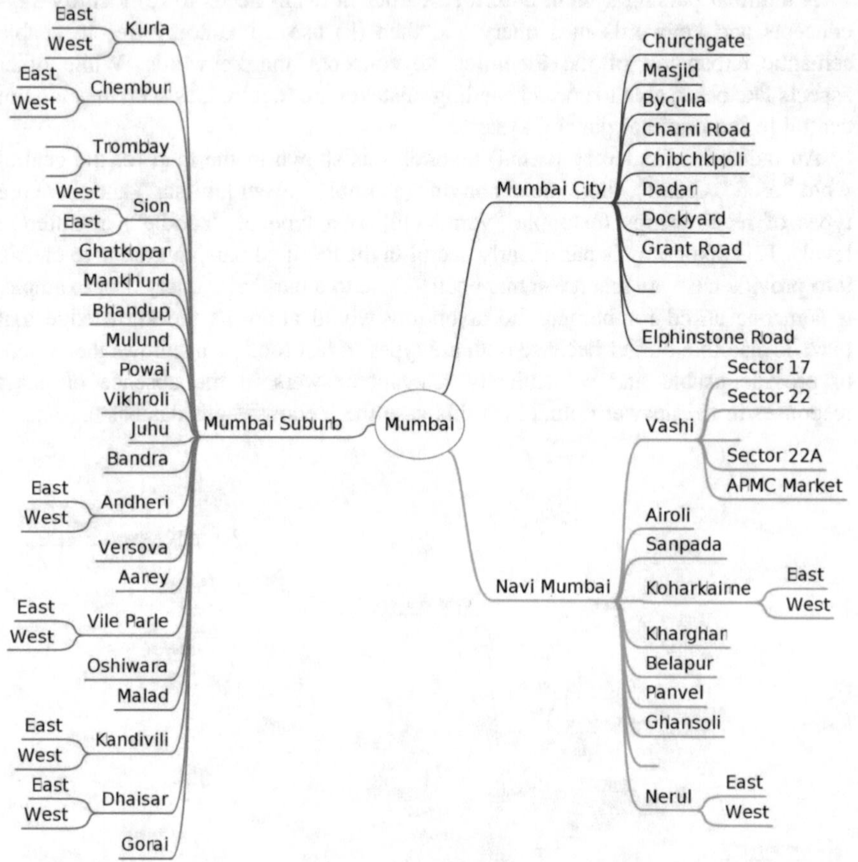

The above chart (central word "Mumbai") shows a partial taxonomy of locations in Mumbai. Most of the locations in suburban regions have a distinct tag of "East" and "West"; for example, Andheri (East) and Andheri (West). The taxonomy gives a broad sense of the physical proximity of one location to another (for example "Juhu" and "Bandra" are adjacent locations and hence appear close to each other in the visualization). However, for obvious reasons, "time required to travel" is also an important criterion for determining proximity. For example, in this case, though Dadar is close to Chembur in terms of distance and adjacency, in terms of travel time, Byculla is much closer to Chembur than Dadar because of the Eastern freeway connectivity between Chembur and Byculla. The minimal parsing system, in the absence of a "DTDC courier in Chembur", would respond as first choice the address of "DTDC courier in Byculla" rather than with the address of "DTDC courier in Dadar" (In the scenario that DTDC courier is not present in Chembur but is present in both Dadar and Byculla). These aspects make the minimal parsing system both intelligent and user friendly—more human like.

The advantage of using this strategy can be seen from the following examples from a working system that was actually operating in the field and accessible and used by large numbers of users for over a year after implementation.

Scenario 1 A user is looking for a "`Studio in Eastern Andheri`".

An ordinary search strategy would fetch:

[1] EASTERN TRADERS, ANDH [2] EASTERN ELECTRONICS [3] M K EST, 73, M K EST, ANDH (E), 28590034,

from the records of the Yellow Pages database.

However, the minimal parsing based strategy would enable extraction of appropriate and exact information, namely,

[1] M K EST, 73, M K EST, ANDH (E), 28590034
[2] GEMINI STUDIOS, C/3, M.I.D.C., ST NO 11, ANDH (E), 28229933
[3] KAMAL AMROHI STUDIO, JOGESHWARI VIKHROLI LINK, ANDH (E), 28208026
[4] CHANDIVALI OUTDOOR STUDIO, CHANDIVALI RD, ANDH (E), 28521097

Note that an ordinary search strategy would produce any results with any of the words in the query ("studio", "eastern" and "Andheri") as the keywords, hence Eastern Electronics and so forth; minimal parsing system would be able to return better search results by understanding that the query intends to search for a "studio" in 'Eastern Andheri' because it is able to associate semantic dimensions to the words "studio" and 'Eastern Andheri".

Scenario 2 For the query "Want to have Meduvada in Juhu"

A keyword-based search engine would not return any results while our system would return

[1] UDIPI SHREE KRISHNA, JUHU CHURCH RD, JUHU, 26713178
Following may also be useful:
[2] THE SEAFARER REST, LIONS, CNTRL JUHU BEACH, JUHU, 26162839
[3] SAPPHIRE, THE EMERALD, JUHU TARA RD, JUHU, 26611150
[4] SUBURBIA REST & BAR, GAYLAND HTL, JUHU TARA RD, JUHU, 26170999

Scenario 3 Suppose a user queries looking for "DTDC Couriers in Sanpada".

In the absence of a DTDC courier in Sanpada, the ordinary search strategy would produce no output, while an NLP-based system would give close and appropriate records extracted from the Yellow Pages directory

[1] DTDC STALLION ENTPS,STALLION ENTPS, A 31, VASHI PLAZA., SEC 17., VASHI, 27894652
[2] DESK TO DESK COURIERS, 7, GR FLR, AMBASSY CENT, NARIMAN POINT, 56311357
[3] DESK TO DESK COURIERS, GALA NO 15, 1ST FLR, MEHTA STATE, NDHERI KURLA RD, ANDH(E), 56943478
[4] DTDC COURIERS, 15, MEHTA EST, AND-KURLA RD, ANDH(E), 56943477

In the absence of a "DTDC courier" in Sanpada, the system is able to dilute the location dimension to search for "DTDC courier" in locations adjoining "Sanpada". The ability to do this is because the minimal parsing system is able to assign a semantic dimension to "Sanpada"; the taxonomy tree is able to gauge all the areas (in the same dimension) close to "Sanpada". So it generates answers regarding DTDC offices in Vashi. Note that "Vashi" is a suburb close to "Sanpada".

Scenario 4 For a query, "Citibank ATM in Vashi"

A keyword-based search strategy returns no results (because there was no Citibank ATM in Vashi at that time). On the other hand, minimal parsing system is able to give information that is useful while suggesting that there is no perfect fit for the query posed. It says

"No perfect fit for CITIBANK ATM IN VASHI. Hope this helps:

[1] CITIBNK, PANCHEEL ARCD, SEC 5, AIROLI5,
[2] INDUSIND BNK LTD, MANEK CPLX, SEC 29, VASHI,
[3] UTI BNK SHP 1, PL 17, SHIV DARSHAN, SEC 4, VASHI
[4] UTI BNK WARDHAMAN CHMBS, PL 84, SEC 17, VASHI, 27660066"

Scenario 5 For the query "Breakfast in Taj"

A keyword-based search would result in no record or any record having Taj as the company name, however minimal parsing system results in correct results because of the ability to relate breakfast to a place which serves food. The records returned are

[1] TAJ GROUP OF HTL, MANDLIK HSE, MANDLIK RD, COLABA, 22022626
[2] TAJ GROUP OF HTLS, MANDLIK RD, APOLLO BUNDER, COLABA, 56653366

[3] TAJ PRESIDENT, 90, G D SOMANI RD, CUFFE PARADE, 56650808
[4] TAJ LANDS END REGENT, LANDS END, BANDSTN, BDRA(W), 5668123

Scenario 6 For a query "Buying Jeans in Andheri"

A keyword search strategy produces no results while minimal parsing system gives

[1] IMAGE APPARELS P LTD, ARVIND CHMBS, WERN EXPRESS HIGHWAY, ANDH(E), 28224892
[2] LIVE IN JEANS, C-6, MIDC, RD NO 22, ANDH(E), 28252127
[3] APEX JEANS WEAR, 9/F, NANDJYOT INDL PREMISES, ANDH KURLA RD, ANDH(E), 28511891
[4] SINGAPORE OLLECTION,DN RD, ANDH(W), 26209109

Scenario 7 For the query "Cable operator in Vashi."

An ordinary search strategy produces no results, while minimal parsing system would give

[1] SSV CABLE P LTD,9, NR INDIAN BNK, LANDMARK CHS, SEC 14, VASHI, 27664073
[2] AASHISH CABLE NET INDIA P LTD, SEC 9 A, GURAV HALL, VASHI, (O)27655535.
In addition it also lists
[3] SEVEN STAR CABLE, SHP-3, MINI JEWEL, OPP GTB BNK, SEVEN B'LOWS, ANDH(W), 26362675 and
[4] UCN HATHWAY CABLE, 2ND FLR, STRAND CNMA BLDG, COLABA, 22812994

suggesting them as possible alternative answers to the query.

These examples demonstrate the value-add derived from using the minimal parsing strategy to retrieve information from the database. As all these real-life queries and response demonstrate, the ordinary search strategy fails because of either the absence of the information in the Yellow Pages directory or the inability to extract more information from the query. The minimal parsing approach, on the other hand, has shown itself to be really useful.

A prototype of the system that can access the Yellow Pages directory for the Mumbai Yellow Pages directory was deployed for a major telecom operator in India in early 2000. The following are actual screenshots from the pilot.

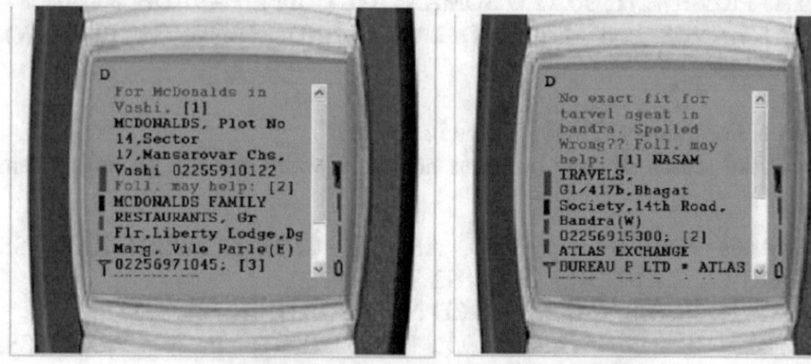

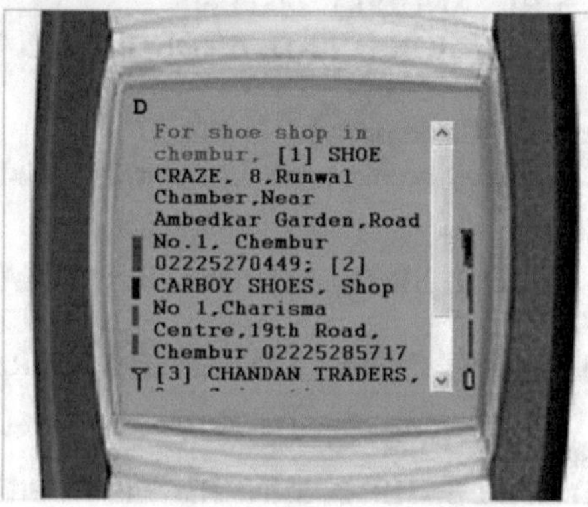

6.5 KisanMitra: A HMI for Rural Indian Farmers

6.5.1 Background

There are several instances where one needs very crucial information but is unable to access it. This can be because, being not well educated, he is unable to frame the query: either because he is unable to convert his thoughts into a syntactically correct query due to lack of formal education or because he is himself not clear about exactly what information he seeks. The Indian farmer falls into this space, especially when he has queries regarding his farm or crop. Clearly, this scenario is best addressed by the minimal parsing system that we have discussed in our earlier chapters. The inability to construct queries which are semantically and syntactically correct leaves the deep parsing approach at the starting block of the race. Given that the information the farmer requires is crucial, it would not be appropriate to leave it to the farmer to pick up the proper answer from a heap of possible answers. This means that the conventional search approach is a non-starter. Minimal parsing approach is the clear winner.

Indian farmers in rural areas not only need expert and timely suggestions to maximize the harvest from their crops; they also need information regarding the subsidies and schemes available from the government to make cultivation pay rich dividends. Conventionally, such guidance comes in the form of a human expert visiting the village now and then; the farmers would assemble and each wait to get their turn to seek answers to their queries. We designed a question answering (QA) system, (named KisanMitra [4], or farmers' friend) to act as a friendly tool which would act as an expert and answer queries of the farmers on a round the clock, round the year basis.

Such a tool not only gives access to information 24 × 7 but is able to keep the information that reaches the farmer updated in real time. It also addresses the farmers' needs in terms of enabling them to query in their own language; laxity with grammar and loose syntax constructs are tolerated. The minimal parsing based system is intelligent in the sense that it understands the intent of the query and provides responses from a well-maintained knowledge-base. In the absence of an exact answer in its information resources, it provides approximate answers which are close to the ideal in some semantic sense.

6.5.2 Introduction

As explained earlier, QA systems can be looked upon as intelligent search engines that can act on natural language queries in comparison to the conventional keyword-based search engines. In general, QA systems accept queries in natural language, extract the intent of the query by parsing the query and then search in their knowledge-base. One of the general requirements of a QA system is to act gracefully by providing close answers from the knowledge-base, in case exact answers do not exist. QA systems are aimed at making them smart and human friendly so that they accept queries in natural language, understand the intent of the query and then respond quickly with answers which are useful and apt.

KisanMitra was built keeping in mind that the rural farmers are the main users. The system covers various issues concerning the farmer in rural India. It is aimed to act like an expert, giving information to queries posed by farmers related to fertilizers, pesticides and government schemes. The system is built on the concept of minimal parsing for reasons mentioned earlier.

KisanMitra is a web-enabled system, with Hindi as the transactional language. The system is capable of correcting spelling mistakes as well as understanding even grammatically incorrect constructs of a sentence. This helps the users to pose their query without any constraints. This minimal parsing technique induced power of KisanMitra allows the user to query the system in the same form as (s)he would articulate to another human to get a response (Fig. 6.12).

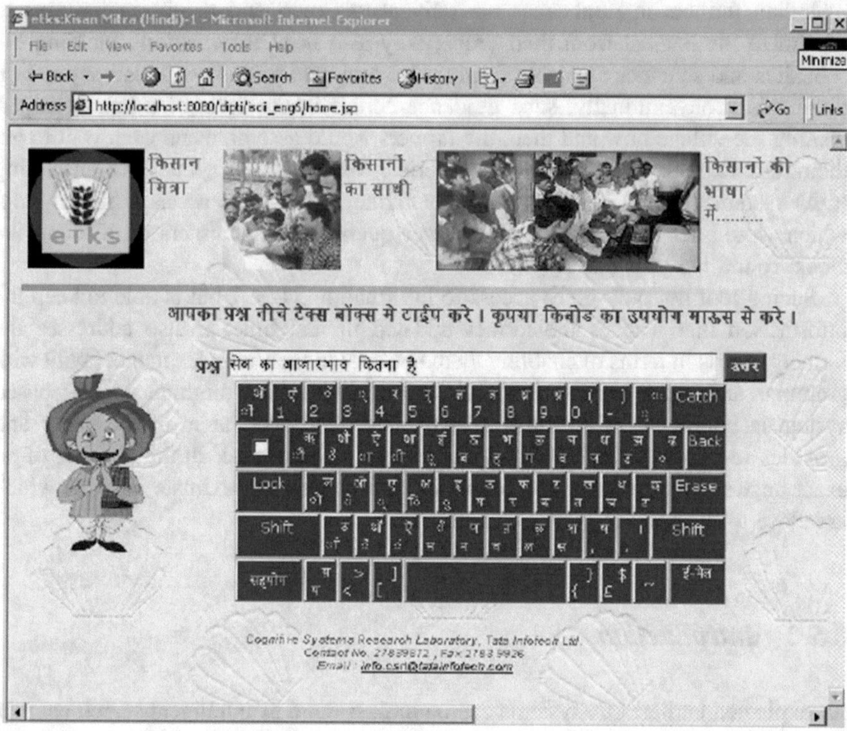

Fig. 6.12 KisanMitra front end. Hindi keyboard is optional and is provided to facilitate easy input in Hindi

6.5.3 *QA Systems as Pattern Matching Systems*

One way to look at QA systems is to think of them as being kc-kw pattern matching systems. The query (as framed by the user) is processed in its original form; it is parameterized and made available to the system in a form that can be used to match the answer paragraphs in the knowledge-base available to the QA system.

In the minimal parsing based system, the pattern can be looked upon as being made up of a keyconcept and associated keywords in the form kc(kw1, kw2, …). Note that for this pattern matching to work, the answer paragraphs in the knowledge-base have also to be processed and parameterized in a similar fashion. This can be done either automatically or manually. The extracted pattern from the farmer query is then matched with the corresponding patterns pre-extracted from the answer paragraphs in the knowledge-base. To allow for more than one possible matches, each matching paragraph is assigned a matching score to quantify the degree of closeness of the query pattern. This permits semantic rank-ordering the answer paragraphs by comparing scores.

In both minimal parsing and full parsing, the matching methodology adopted is the same: namely extraction of patterns from both the query and the answer paragraphs and then matching them so as to find the best answer pattern that matches the query pattern. The difference between the two is in the manner in which the text is processed. The kind of text processing would generally depend on the type of patterns that need to be extracted. For instance, in a conventional keyword type system, the parameter extraction would involve removal of all stop words (example, punctuation marks, connecting words, pronouns, adjectives, etc.) and what remains form the match pattern. For a full parsing system, it would retain the punctuations to help verify the syntactic and semantic correctness of the query before extracting a pattern.

Most QA systems take the approach of full parsing to comprehend the query. While this is good in principle, it is far from satisfactory in practice, in the sense of performance. This is because of the following.

1. The availability of a reliable parser for the language in which the query is input
2. Separate parsers being required for each new language to be used by the QA system
3. The questioners as well as the persons preparing the answer database have all to be well conversant with the rules of grammar of the language of interaction.

If any of these conditions fail, the QA system will not perform properly. Even after one takes care that the parser follows the rules of grammar, the parser could run into problems if the person inputs a query that is grammatically incorrect. It is much too constraining (and hence impractical) to expect a high enough degree of grammatical competence from a casual user (e.g. farmer) of the system. For example, there is a good chance that a full sentence parser would parse a (grammatically) incorrect query from a user and assign a different sense to the query (for example "I sell wheat where for maximum rupees" would not be parsed[1] at all while the equivalent grammatically right sentence "Where can I sell wheat for maximum profits" can be parsed). The subsequent analysis (pattern building) based on such a (wrong) interpretation of the query by the system could in some cases be far from the actual intent of the query. In addition, it is well known that even the best parsing algorithms have their shortcomings and do not necessarily operate correctly all the time!

In the context of KisanMitra, because of the type of intended user, the power of full parsing is blunted because questioners would in all likelihood use casual/oral grammar (to convey intent without worrying about the intricacies of grammar).

Another important point to remember is that, considering the linguistic diversity on the Indian scene, a facility like KisanMitra would be expected to operate with several languages. Implementing full parsers for all these languages will multiply the effort of implementation. A minimal parsing approach, being minimally grammar dependent, makes implementation across languages much simpler.

[1]http://www.link.cs.cmu.edu/cgi-bin/link/construct-page-4.cgi#submit.

6.5.4 Our Approach

The goal of a QA system in the KisanMitra context is therefore to provide a correct answer wherever it is possible and, in case a correct answer is not available, to give an approximately correct answer in response. This makes the system robust and usable. The minimal parsing system, based on extraction of keyconcepts and keywords from the query using a taxonomy tree is best suited for this. The complete architecture is captured in Fig. 6.13.

The KisanMitra architecture allows the user to input his query in Hindi; the root of the input word is first extracted (for example the root form of फसलों is फसल) and the root word is passed through a Hindi to English lookup dictionary (example फसल → crop) to translate the Hindi query word by word. Note that word by word translation (because of word order and other differences between English and Hindi) is certain to produce a syntactically erroneous English sentence.

For example, the query

"गेहूँ के फसलों की कटाई कैसे की जाती है?"

(gehoon ke phasalon kee kataee kaise kee jaatee hai?)

should actually translate to

"`How is wheat harvesting done?`".

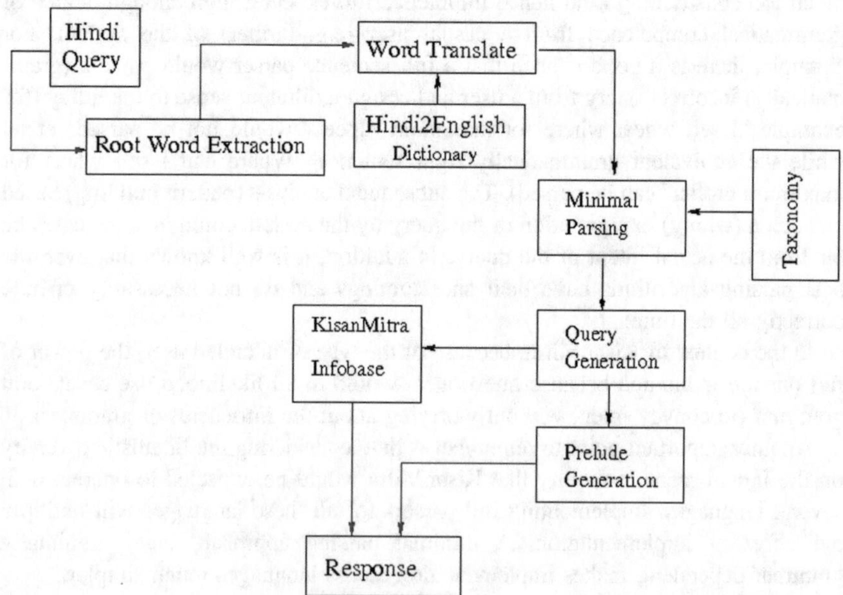

Fig. 6.13 Functional architecture of KisanMitra. Note that the blue lines indicate the text in Hindi while the black lines represent English text usage

However, word by word translation would result in

"`Wheat of crop of harvest how done`"

This preserves the semantics but is far from being syntactically correct. This "sequence of words" cannot be processed by a full parsing system but is easily taken care of using the minimal parsing system described earlier in this chapter. The resultant expression would be `harvest(wheat, crop, how)`.

The minimal parsing block extracts kc's and kw's from the query making use of the taxonomy tree. As mentioned earlier, the taxonomy is generally constructed or engineered manually for a given domain and superimposed on a generic taxonomy (which is domain agnostic) when applicable. Typical taxonomy (see a sample sparse taxonomy, for this domain, below) can be thought of as a special kind of a WordNet that holds the relationships between different words. Additionally, each word in the taxonomy has a kw, kc or qt tag corresponding to keyword, keyconcept and question type, respectively. These tags are generally derived through statistical analysis of the domain data for which the QA is being built. For example, a typical entry in the taxonomy for a word tagged as kc looks like

Root: `procedure`
Syn: `steps, ways, method, process`
Type: `how`
Part: `instructions`
Class: `kc`

where "`procedure`" is the word (denoted by Root: `procedure`) which is a keyconcept (denoted by Class: `kc`). The words "`steps, ways, method, process`" are the *synonyms* (denoted by Syn: `steps, ways, method, process`) and the word "how" (denoted by Type: `how`) is a hyponym (*type of*) the word "`procedure`" while the word "`instructions`" is a meronym (*part of*) the word `procedure` (denoted by Part: `instructions`). It is to be noted that the construction of the taxonomy makes use of the words and phrases that the user of the system is likely to use and not based on the strict meaning of the word. For example, the word "`tips`" is not likely to be categorized as a hyponym of the word "`definition`" in the strict sense of usage or English grammar but in practice, especially in a transactional scenario, it is used as a hyponym of the word "`definition`".

A sample taxonomy is shown in the table below.

S. No.	Root	*Synonym*	*Hyponym (Type of)*	*Meronym (Part of)*
1	where(qt)			
2	benefit(kc)	reason, use, utilization, why, utility		
3	definition (qt)	what, describe, brief, detail, information, data, list, name, tell, provide, give, know	tips	
4	procedure (kc)	steps, ways, method, process	how	instructions
5	fruit(kw)		musk_melon, water_melon, mango, grape, guava	peel
6	day(kw)		week_day, week_end	
7	when(kw)	how_long		
8	year(kw)			
9	month(kw)		January, May, February, March, April, June, July, August, September, October, November, December	
10	quantity (kc)	number, count, amount, size	how_much, how_many	
11	how_much (qt)			
12	crop(kw)		oil_seed, wheat, ground_nut, maize, rice, cotton, sugar_cane, gram, turmeric, potato, onion, capsicum, tomato, spinach, chillies, garlic, radish, cabbage, jowar, carrot, cauliflower, gourd, cucumber, tar, beet_root, barley, turnip, paddy	root, flower, fruit, seed, leaf, grain, branch, stem
13	apply(kc)			
14	availability (kc)	how_to_get, can_get, where_can_get, accessibility		
15	effect(kc)	consequence	fallout, result, repercussion	
16	importance (kc)	significance, prominence		
17	retain(kc)	remain_with, hold, keep		
18	type(kc)	kind, different, variety		
19	attract(kc)	pull, draw, appeal		

The minimal parsing system triggers the query generation module; the generated patterns are sent to the KisanMitra knowledge-base to fetch answers that have patterns that match the question pattern. For example, for the query,

"`How can I harvest my moong dal which is infected by pest_X`",

the query generating module would generate a

```
keyconcept (keyword1, keyword2,...)
```

pattern. In this case, the pattern `harvest(pest_X, moong_dal)` is generated; if no answer with this pattern were to be found in the knowledge-base then it would generate the pattern `harvest(pest_X, pulses)`, where `pulses` is a hypernym of `moong_dal`; This is followed by `harvest(pest_Y, moong_dal) where pest_Y` is a co-hyponym of the word `pest_X` followed by `harvest (pest_Y, pulses)`. Clearly, all these patterns fetch answers that are close in some sense to the original query, namely, `harvest(pest_X, moong_dal)`, posed by the farmer.

Additionally, the query generation module triggers the prelude generation module, which is active especially when the patterns extracted from the query and the pattern "searched" for in the knowledge-base is different. For example, the options `harvest(pest_X, pulses)`, `harvest(pest_Y, moong_dal)` and `harvest(pest_Y, pulses)` have a prelude (generated by the prelude generation block) affixed to the response because these are not exact responses to the query of the user but are only "close" to the actual query.

The generated prelude, (e.g. "आपने मूंग दाल को कैसे काटने के लिए पूछा, लेकिन मैं दालों को काटने के बारे में बताऊंगा।", i.e. "You asked for how to cut moong dal, but I will tell about cutting pulses." in English) if any and the answer paragraph (in Hindi) are output as a response to the user's Hindi query. In Fig. 6.13, the Hindi text processing is shown in blue while the English text processing is shown in black. Note the smooth functioning of the minimal parsing system in Hindi as well as in English. It can easily be seen that the system can be extended in a straightforward manner to work in any other language.

As mentioned above, note that the minimal parsing system would be able to process the syntactically incorrect (word translated) English query "`wheat of crop of harvest how of done`" would result in the pattern "`harvest (wheat)`". This is the advantage of using the minimal parsing system, especially because building a language translation system (Hindi → English) is much harder and prone to errors compared to word translation which is essentially a word look up in a Hindi → English dictionary.

Thus, one of the main advantages of using minimal parsing system to answer transactional queries is that any functional system in English can be easily re-tooled to work for any other language (say Malayalam). As seen earlier the only resource that is required is the word Malayalam to English dictionary. There is no specific need to build a separate taxonomy in Malayalam (see Fig. 6.13).

References

1. https://en.wikipedia.org/wiki/Sentience
2. S. Kopparapu, A. Srivastava, S. Das, R. Sinha, M. Orkey, V. Gupta, J. Maheswary, P.V.S. Rao, Accessing yellow pages directory intelligently on a mobile phone using SMS, in *MobiComNet 2004*, Vellore
3. https://developers.google.com/maps/documentation/distance-matrix/
4. S. Kopparapu, A. Srivastava, P. Rao, Kisanmitra: a question answering system for rural indian farmers, in *International Conference on Emerging Applications of IT (EAIT 2006) Science City Kolkata*, February 10–11, 2006 (2006)

Chapter 7
Conclusions

While natural language is complex and difficult for machines to interpret, it is not impossible to actually build usable and human-friendly natural language human–machine interfaces (HMIs) for use by masses. This monograph while dwelling into the complexities of language and grammar and capturing aspects that makes machine process natural language difficult also tries to show that there exists simple techniques and mechanisms which can make machine processing of natural language possible in a certain context of what is called transactional interactions.

The monograph gives several examples of HMI in varied areas from Yellow Pages to book to a self help system for non-English language. These interfaces are enabled through an underlying concept of minimal parsing which is a simple yet powerful technique that makes understanding the intent of the query with a fair degree of confidence. Also, the underlying architecture of the HMI system contributes to the overall system performing as well as a human–human interaction. The main distinction being that the system based on minimal parsing is able to give feel answers rather than a logical think answer.

The minimal parsing system while exploiting the constrained domain and the limited number of ways in which queries of transactional grammar in nature can be asked by human works by analysing the natural language query in a minimal manner rather than adopting a conventional deep parsing approach. The minimal parsing approach also assumes that humans by nature are not necessarily grammatically correct or consistent especially when trying to transact to obtain some information. For this reason, minimal parsing systems flourish when the more conventional deep parsing methods fail.

In this monograph, based on some of the implemented work done by us in the past, we give some insights into the evolution of language and demonstrate the richness of the languages which in turn makes the lexical words have several senses which in turn makes understanding (even by humans) difficult. We speak about the work that broadly addresses the natural language understanding from the angle of a question answering system. We deliberate on the use of deep parsing for natural language query understanding from both the standpoint of the complexity of the

© Springer Nature Singapore Pte Ltd. 2018

P. V. S. Rao and S. K. Kopparapu, *Friendly Interfaces Between Humans and Machines*, https://doi.org/10.1007/978-981-13-1750-7_7

language and also the inability of the user of the system to be grammatically correct all the time on one side and also deliberate on the other side the inability of keyword-based search system. We then weave a middle path which we call minimal parsing. We argue that minimal parsing approach allows the interfaces to be more usable and human friendly.

Appendix A

A.1 Example Taxonomy Tree

P. V. S. Rao, SunilKumar Kopparapu

This taxonomy tree was built for a well know book on Information Technology. Note that a row like

method (kc)	technique, procedure, way	top_down, modularization, conventional	process, how

Represents that the word "method" is a keyconcept and the words "technique", "procedure" and "way" are synonyms while top_down, modularization, conventional have a Type of relationship with the word "method" while the words "process" and "how" have a Part of relationship with the word "method". It should be noted that there are several words in the taxonomy that are abbreviations (see for example the entry 'oltp' in the second row means 'online transaction processing') associated with domain. The idea of presenting this 'actual' taxonomy, which we handcrafted, is to give a feel to the reader the kind of taxonomy tree that they might have to build to enable a minimal parsing system.

Root	Syn	Type	Part
method (kc)	technique, procedure, way	top_down, modularization, conventional	process, how
process (kw)		oltp, routine	
conventional (kw)	traditional		
how (qt)			
process_specification (kw)		non_procedural, procedural	clause
non_procedural (kw)		dt	
dt (kw)		eedt, ledt, erdt, medt, ldt	decision_rule, dash_entry, dont_care_entry

(continued)

© Springer Nature Singapore Pte Ltd. 2018
P. V. S. Rao and S. K. Kopparapu, *Friendly Interfaces Between Humans and Machines*, https://doi.org/10.1007/978-981-13-1750-7

(continued)

Root	Syn	Type	Part
dash_entry (kw)	dash		
procedural (kw)		se	
se (kw)			
clause (kw)	condition	action_clause, connectivity_clause	
minimization (kc)	reduction	removal, merge	
removal (kc)	elimination, eradication		
merge (kc)	combination, integration, join, blend, mix, unite		
project (kw)			sdlc
sdlc (kw)			development, design, analysis, proposal, implementation, test
development (kc)	build, creation, construction, formulation, preparation		
design (kc)	modelling, layout, formalization, authoring	oom	draw, plan
oom (kw)		uml	polymorphism, inheritance, abstraction, encapsulation
uml (kw)			class_diagram, crc, state
class_diagram (kc)			class, object_diagram
object_diagram (kw)	instance_diagram		object
object (kw)	instance	tangible_object, intangible_object	
crc (kw)			crc_team, class_index_card
inheritance (kw)		generalization, specialization	subclass, superclass
encapsulation (kw)	information_hiding		
draw (kc)	sketch		
plan (kc)	strategy	test_plan, implementation_plan, review_plan	
analysis (kc)	study	feasibility_analysis, er_analysis	srs
feasibility_analysis (kw)			feasibility, cost_benefit_analysis, payback
feasibility (kw)		operational, technical, economical	
cost_benefit_analysis (kw)			cost, benefit, shortcoming
cost (kw)		direct, indirect	
benefit (kc)	advantage, merit, preference	tangible, intangible, direct_saving	
shortcoming (kc)	demerit, disadvantage, pitfall, drawback, flaw, limitation, constraint, restriction, deficiency	problem	
problem (kc)	difficulty, obstacle, complexity		
payback (kw)		simple_payback, present_value	
er_analysis (kc)			er_model, normalization, unnormalization
er_model (kw)	er_diagram		entity_set, relationship_set
entity_set (kw)			entity

(continued)

(continued)

Root	Syn	Type	Part
entity (kw)		external_entity	
relationship_set (kw)			relationship
relationship (kw)		1_n_relationship, cardinality, participation	
normalization (kc)			successive_normal_forms
successive_normal_forms (kw)			nf
nf (kw)		1nf, 2nf, 3nf, 4nf, 5nf, bcnf	
unnormalization (kc)			
participation (kc)		compulsory, conditional, optional	
compulsory (kw)	mandatory		
srs (kw)			specification, prioritization
prioritization (kc)			
specification (kc)		final, multiple	
multiple (kw)	many, more	two, three, four, five	
proposal (kw)	system_proposal		
implementation (kc)	realization, accomplishment, execution		
test (kc)		parallel_run, program_test, pilot_test, string_test, alpha_test, system_test	
tool (kw)	aid	diagram, model, k_map	
diagram (kw)	figure, picture	document_flow_diagram, context_diagram, dfd, block_diagram, flowchart, symbol	
document_flow_diagram (kw)			dash_line, solid_line, dash_arrow
dfd (kw)		top_level, logical_dfd, physical_dfd	sink, data_store, data_flow
symbol (kc)	sign, notation, terminology		circle, arrow, rectangle, diamond, parallel_lines
circle (kw)	bubble		
model (kc)		data_model, physical_model, object_oriented	
data_model (kw)		conceptual_data_model, logical_data_model, external_data_model	
k_map (kw)			variable, dont_care_entry, adjacent_square
variable (kw)		two_variable, three_variable	
adjacent_square (kw)		two_adjacent_square, four_adjacent_square	
communication (kw)		message, interview, discussion, questionnaire	interpersonal_relations, interaction
discussion (kw)		group_discussion	participate, decision
participate (kc)			
decision (kc)		consensus	choice
choice (kc)	selection, pick		
interview (kw)			question

(continued)

(continued)

Root	Syn	Type	Part
network (kw)		physical_network, logical_network, van, lan, pstn, internet, intranet, extranet, logical_layer	modem, bridge, router, private_leased_lines, protocol, packet_switching, system, security, isp
internet (kw)			www
www (kw)			web_site, web_browser, web_server, url
web_site (kw)		search_engine	web_page
web_page (kw)			markup_language
markup_language (kw)		sgml	tags
sgml (kw)			html, xml
pstn (kw)	pots		
protocol (kw)		http, ftp, mime, smtp, set_protocol, netbill_protocol, ip	
ip (kw)			ip_address
ip_address (kw)			domain
packet_switching (kw)			packet
packet (kw)			header
system (kw)		computer_system, information_system	maintenance, recovery
computer_system (kw)			computer
computer (kw)	pc		h_w, s_w, output, input
h_w (kw)		device	
device (kw)		output_device	
output_device (kw)		printer, vdu, audio_output_unit, screen	
printer (kw)		impact_printer, non_impact_printer, laser_printer, inkjet_printer, line_printer, character_printer	print
s_w (kw)		virus	program
virus (kw)	computer_virus	boot_sector_virus	
program (kw)		edit_program, control_total, application_program, data_validation_program	indentation, loop, algorithm, code, statement, hll
loop (kw)	repetition	for_loop, while_loop	
for_loop (kw)	for_structure, for_statement		
while_loop (kw)	while_structure, while_statement		
code (kw)		serial_numbers, block_codes, group_classification_code, significant_codes, modulus_n_code	
modulus_n_code (kw)		modulus_11_code	weight
statement (kw)	sentence	decision_structure, imperative_statement	
decision_structure (kw)			conditional_statement
conditional_statement (kw)		if_else, if, if_then_else, case, default	
information_system (kw)		dbms, dss, mis, e_commerce, example_system	data

(continued)

(continued)

Root	Syn	Type	Part
example_system (kw)		enquiry, vendor_supply, billing, journal_processing, store	
store (kw)	shop	virtual_shop	
dbms (kw)		rdbms	database
database (kw)		file, stock_ledger	table, data_dictionary
file (kw)		master_file, data_file	file_header
table (kw)	relation, schema		tuple, dependency, subschema
dependency (kw)		functional_dependency	
data (kw)		information, fact, opinion	character, range, value
information (kw)	processed_data	strategic_information, tactical_information, operational_information, statutory_information	
quantitative (kc)	volume	less, small, large	
output (kc)	result, affect, effect, consequence, layout	display	output_element
display (kc)		gui, business_graphics	
business_graphics (kw)			color, map, graph, chart
map (kc)			
chart (kw)		print_chart, pie_chart, bar_chart	
input (kc)	feed, entry	data_entry, interactive	input_element, data_element
data_entry (kw)	data_input	offline, online, batch, form	
data_element (kw)	data_items		
form (kw)		feedback	
interactive (kw)		menu, template, command	
e_commerce (kw)		netbill	transaction, payment, edi
payment (kc)	money	credit_card, e_cheque, e_cash, interest, discount, finance	clearance
e_cash (kw)		digital_coin	transaction_blinding
clearance (kc)			
transaction (kc)		business_transaction, e_transaction	
edi (kw)			edi_message, edi_data, edi_standard, transaction_set, exchange
exchange (kc)	interchange		
maintenance (kc)	preservation		
recovery (kc)			
journal_processing (kw)	journal_acquisition		journal
security (kc)	protection, safety, privacy	digital_signature, firewall, cryptography	prevention, password, certification
digital_signature (kw)			hash_function, signing
signing (kc)			
certification (kc)	authentication, authorization		certification_authority
firewall (kw)		packet_screening, filtering_router, proxy_application_gateway, hardened_firewall_host	

(continued)

(continued)

Root	Syn	Type	Part
cryptography (kc)	encode	des, rsa	encryption, decryption, plain_text, ciphertext, symmetry
encryption (kc)	encode		transposition, substitution
ciphertext (kw)	cryptogram		
symmetry (kc)			
prevention (kc)	avoidance, precaution		
data_activities (kw)		mining, archival	
mining (kc)		data_mining	
archival (kc)		data_archive	
operation (kc)	computation	non_numeric, numeric, logical, relational	operator, expression
non_numeric (kw)	non_arithmetic, alphabetical		
numeric (kw)	arithmetic		
expression (kc)	formula	boolean_expression	
boolean_expression (kw)	boolean_algebra		
operator (kw)	identity		
function (kc)	job, task, do, work, activity, role, responsibility, manage, handles, looks_after, takes_care, looks_into, duty	service	goal
service (kc)	help, provision		
goal (kc)	objective, purpose, aim, motivation	sub_goal, main_goal	use, requirement
use (kc)	utilization, application, require, necessity, need, appropriate		
requirement (kc)			
group (kw)	team	people	
people (kw)	persons	manager, computer_professional, user, senior_scientific_officer, scientific_assistant, auditor	who
user (kw)		customer, operational_staff	
manager (kc)	incharge, executive	middle_level_manager, top_level_manager, line_manager	
top_level_manager (kw)		ceo, chairman	
computer_professional (kw)	computer_programmer	analyst, hacker, dba	mind
mind (kw)		analytical_mind	knowledge
knowledge (kc)	awareness, familiar, understanding		
who (qt)	whom		
organization (kw)	establishment, center	industry, institute, fstc, company, bank, office, hostel	department, mess, branch
institute (kw)	university, college	ncsi	campus
industry (kw)		manufacturing	
mess (kw)	canteen, cafeteria, pantry		
department (kw)	section		

(continued)

(continued)

Root	Syn	Type	Part
modification (kc)	change, alteration, updation, upgradation, edit	conversion	
conversion (kc)	transformation		
quality (kc)	skill	good, bad, redundant, traceable, independence, static, non_static	
static (kw)	constant		
non_static (kw)	dynamic		
independence (kc)		data_independence	
good (kw)		accurate, successful, complete, trustworthy, timely, up_to_date, relevant, significant, fault_tolerant, reasonableness, integrity, consistent, right, comprehensive, expandable	
expandable (kw)	flexible		
bad (kw)		delay, ambiguity, incomplete, inconsistent, invalid, irrelevant	
ambiguity (kc)		real_ambiguity, logically_impossible_ambiguity	
redundant (kc)			
accurate (kw)	unambiguous, concise, exact, perfect, ideal, precise		
invalid (kw)	incorrect, wrong, illegal, improper, unauthorize		
integrity (kc)			
consistent (kc)			
explanation (qt)	description, note, narration, detail, overview, introduction, summarization, concept, theory, definition, what, meaning		distinguish, scope, example, interpretation
distinguish (qt)	difference, distinction, not_same, vary, comparison, separate	contradiction	
scope (qt)			
example (qt)	illustration		
dummy2(kc)		levelling, decomposition	
levelling (kc)			
decomposition (kc)	division, break_down		
reason (qt)		why	
why (qt)			
key (kw)		er_key, security_key	
er_key (kw)		composite_key, relation_key	
security_key (kw)		public_key, secret_key	
secret_key (kw)	private_key, symmetric_key		
correction (kc)	validation, legal		check, detection
detection (kc)	search, find, hunt, locate		identification
check (kc)	verification, ensure, clarification, assurance	radix, interfield_relationship	control, total, audit

(continued)

(continued)

Root	Syn	Type	Part
control (kc)		management_control, sequence_numbering, batch_control_record, data_entry_verification_control, check_digit, proof_figure, two_way_check, relationship_check, checkpoint, restart	
total (kw)	sum, addition	record_total	
audit (kc)			audit_trail, audit_package
report (kc)		feasibility_report, manual, executive_summary	heading, report_group, report_generator
manual (kw)		users_manual, system_manual	
heading (kc)			
report_group (kw)			report_heading, page_heading, control_group, control_footing, details_heading, page_footing, report_footing, final_control_footing, detail_line
guideline (kc)	suggestion, convention, point	standard	
standard (kc)			
representation (kc)	depiction, indication, show, express	presentation	format, structure
presentation (kc)			
format (kc)	pattern, syntax		
structure (kc)	architecture, hierarchical_chart, hierarchy	pyramid, layered_architecture	
ability (kc)	capability		
commerce (kw)			business
technology (kc)	trend		
level (kc)			
storage (kw)		disk, tape, cd	
type (kc)	classification, category, kinds, medium, means, mode, different, various		
request (kc)		reorder	
step (kc)	stage, phase, measure, action		
determination (kc)			
availability (kc)	get, gather, collection, derivation, acquisition, achieve, procurement, retrieval, obtain, arrival		
issue (kc)			
attribute (kc)	feature, characteristic, property, aspect		
important (kc)	significant, essential		
source (kc)			
identification (kc)	recognition		

(continued)

(continued)

Root	Syn	Type	Part
reorganization (kc)			
examination (kc)	exploration		
solution (kc)		alternative_solution, proposed_solution	
evaluation (kc)	assessment		
absent (kw)	not_present, not, missing		
rule (kc)	principle	elementary_rule, impossible_rule	
record (kc)		data_record	
dummy (kw)		around, with, through	
sequence (kc)	order, numbering	descending, ascending	
share (kc)			common
similar (kw)	same, synonym, analogue		
establish (kc)	start, begin		
interface (kc)			
observation (kc)			
assignment (kc)			
spread (kc)	transmission, propagation		
expansion (kc)	acronym, full_form, abbreviation, stand		
period (kw)	time, duration, frequency		
partnership (kc)			
error (kc)	mistake, fault	transcription_error, transposition_error	
management (kw)		hrm, production_management, materials_management, marketing_management, finance_management, research_management, design_management, development_management	
possibility (kc)	probability, allowable, permissible		
element (kc)	component, content, part, subset, consist, comprise		

A.2 Sample HMI (e-Book on Fitness)

Here we list a set of actual queries asked by users, during a pilot run, and the responses provided by the minimal parsing based system.

What are the advantages of learning a new sport

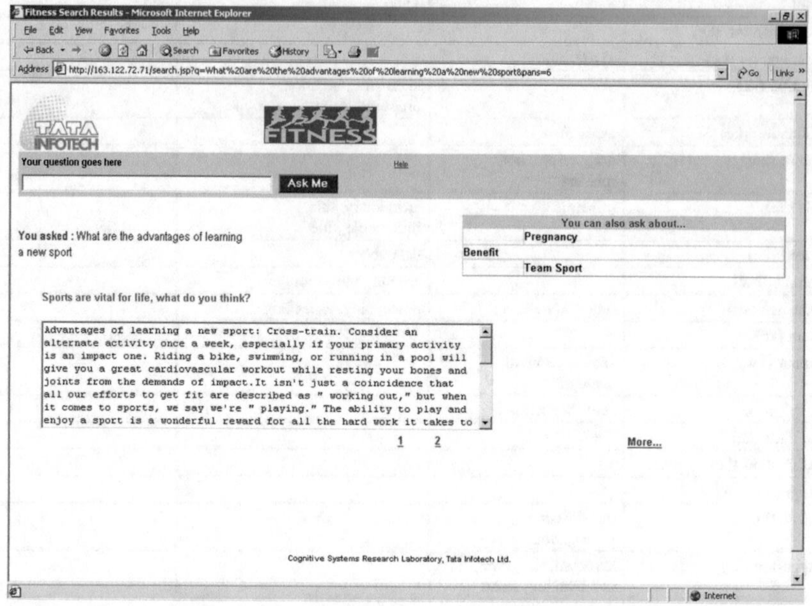

Do you have a method to find my daily calorie needs

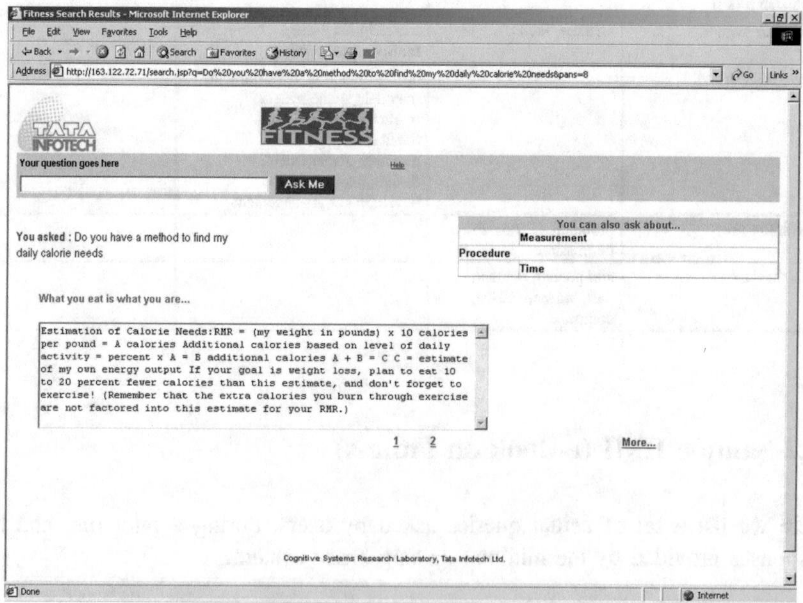

(continued)

(continued)

Somehow a fat-free diet is not working for me. why?

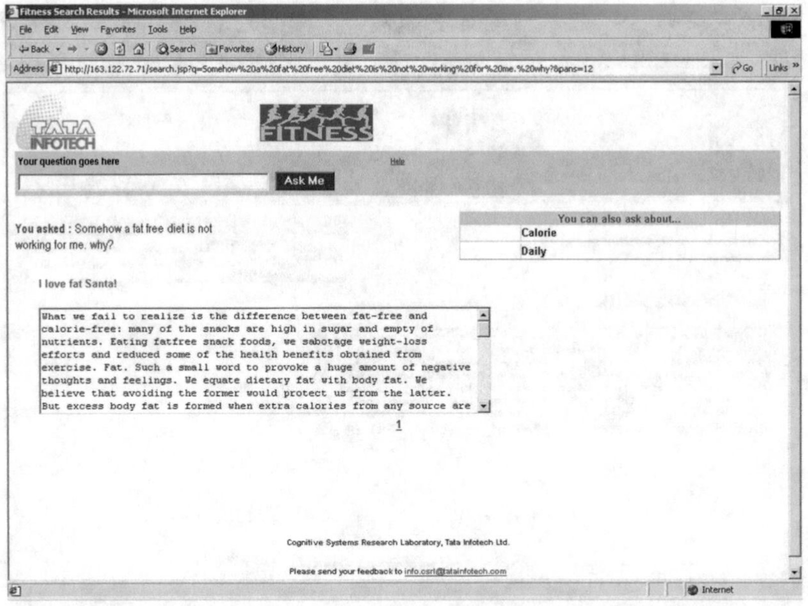

Brief me about some mistakes made by exercisers

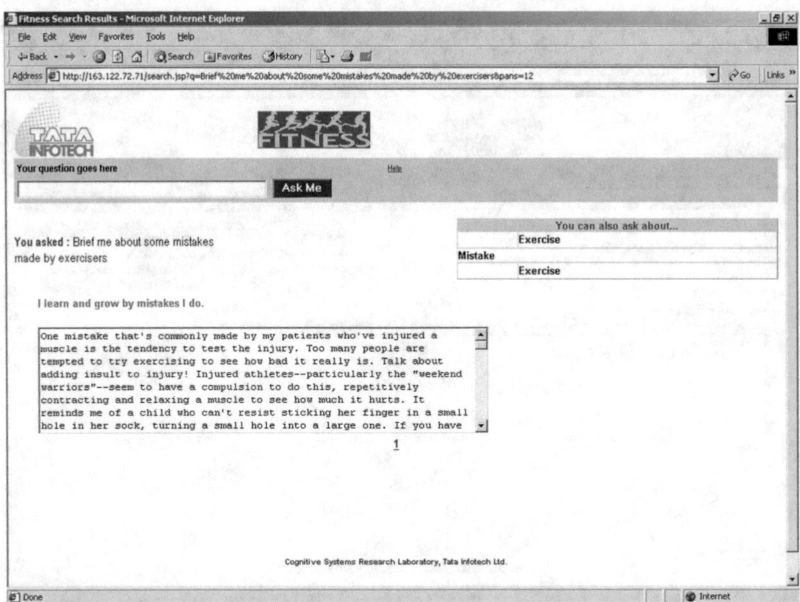

(continued)

(continued)

Can you provide me with a treatment for muscle cramps

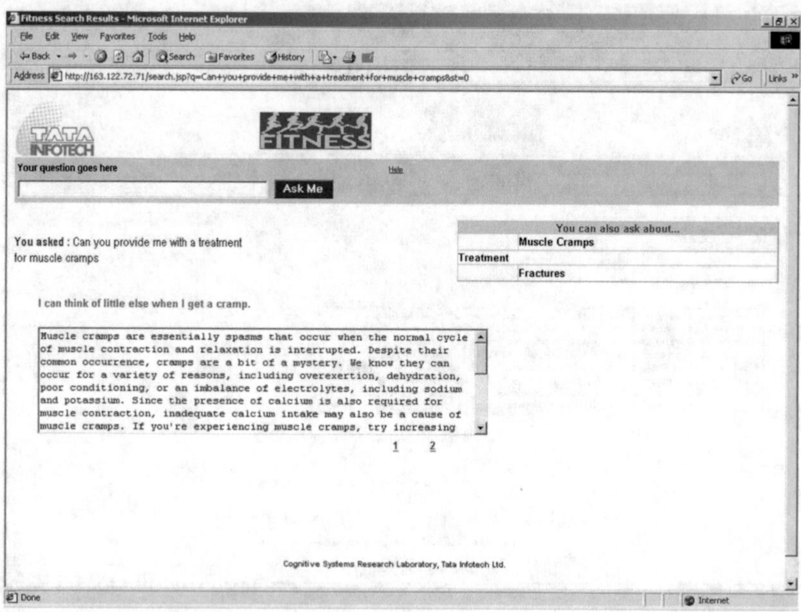

Do you have a cure for stress fractures

Fitness Search Results - Microsoft Internet Explorer

File Edit View Favorites Tools Help

Back Search Favorites History

Address http://163.122.72.71/search.jsp?q=Do%20you%20have%20a%20cure%20for%20stress%20fractures&ans=18

Go Links

TATA INFOTECH

FITNESS

Your question goes here

Help

Ask Me

You can also ask about...

Muscle Cramps

Treatment

Fractures

You asked : Do you have a cure for stress fractures

I get everyone to sign on my plaster.

You seem interested in treatment of stress fractures , I have found details regarding treatment of fractures

Treatment for fractures.
The information will be added soon.

1 2

Cognitive Systems Research Laboratory, Tata Infotech Ltd.

Internet

(continued)

(continued)

How can i exercise in cold climate

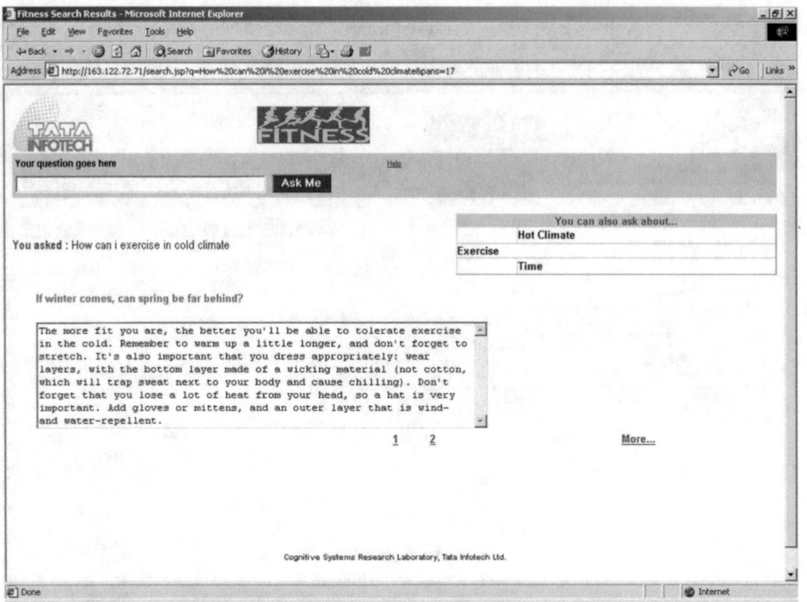

How can i exercise in hot climate

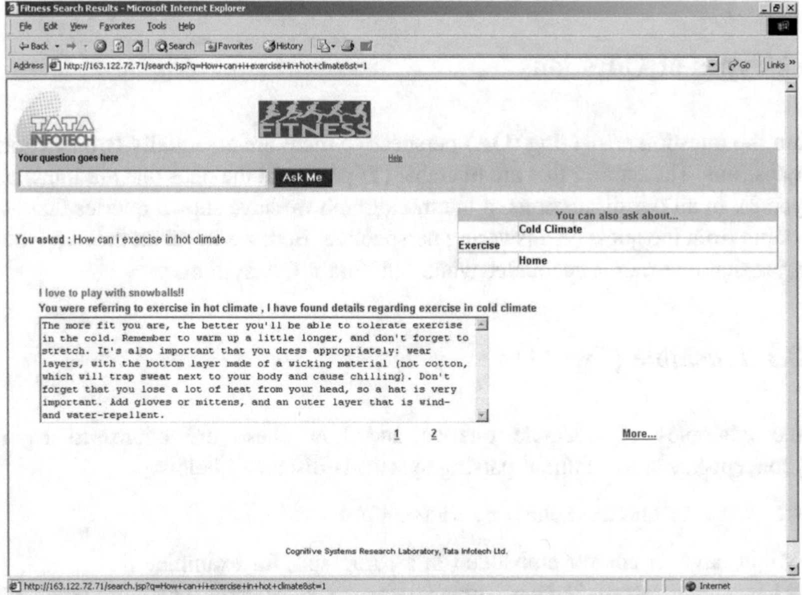

(continued)

(continued)

what fruits should I eat

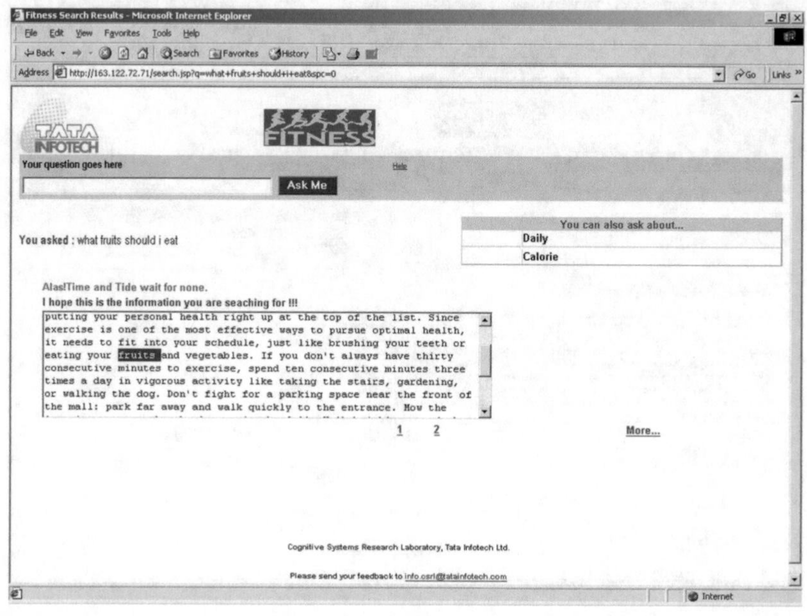

A.3 Types of Questions

From the question answering (QA) perspective there are essentially two categories of questions. The queries that are tractable (Type 1) and the ones that are intractable (Type 2). In all our discussions in the monograph we have shown queries that were tractable from the question answering perspective. Here we dwell on the two type of queries that one might encounter when building a QA system.

A.3.1 Tractable (Type 1)

Some examples of tractable queries and how these are addressed by the keyconcept-keyword minimal parsing system is discussed below.

Q1. What is the nationality of Pope John Paul II?

Might have an answer embedded in a paragraph, for example

P1. "…. Stabilize the country with its help, the Catholic hierarchy stoutly held out pluralism, in large part at the urging of Polish-born Pope John Paul II. When the Pope emphatically defended Solidarity Trade Union during a 1987 tour of the..."

Though not explicitly stated, one can derive the answer from the paragraph P1.

In the notation of keyconcept-keyword minimal parsing system Q1 would look like

```
Nationality(Person = Pope John Paul II)
```

Q2. Which is the fastest car in the world?

```
which(car(fastest), world)
```

Correct Answer: The xxxx, at xxx Km per hour is the fastest car in the world

Wrong Answer: The yyyy will challenge yyyy's lead in the world's fastest growing vehicle market

In the keyconcept-keyword minimal parsing system, based on the dimensionality of the expected answer (in this case "km per hour" or "meters per sec" etc.) one could distinguish the correct answer paragraph from the wrong answer paragraph.

Q3. Who shot Billie the Kid?

```
?shot(Billie the kid)
```

Correct Answer: In 1881, outlaw William H Boney Jr., alias Billy the Kid, was shot by Sheriff Pat Garrett in Fort Summer, NM)

While the answer is right it is not very straight forward because of the alias, namely William H Boney Jr and Billy the Kid being equated to mean the same person, nevertheless a keyconcept-keyword minimal parsing system should be able to identify this as the right paragraph.

Wrong Answer: The scene called for Phillip's character to be saved from a lynching mob when Billy the Kid (Emilio Esteviz) shot the rope in half just as he was about to be hanged.

Notice that there are at least two cues indicating that the answer is wrong.

A.3.2 Intractable (Type 2)

Will India lower the excise tax rates in this year?

Requires general worldly knowledge …
and much more…

Bibliography

1. S. Kopparapu, A. Srivastava, P. Rao, A natural language interface for a railway website, in *Second National Conference on Innovations in Information and Communication Technology 2006*, PSG College of Technology, Coimbatore, 7–8 July 2006
2. S. Kopparapu, N. Janardhan, A novel mobile interface to register citizens complaint, in *iHCI IADIS International Conference Interfaces and Human Computer Interaction 2008*, Amsterdam, Netherlands, 25–27 July 2008
3. P.V.S. Rao, A no-parsing approach to man-machine interaction, in *Proceedings of Symposium on modeling and Shallow Parsing of Indian Languages*, Apr 2006, pp. 242–251
4. Coremetrics [Online]. Available: http://www.coremetrics.com/solutions/onsitesearch.html
5. Indian Railway, Indian rail [Online]. Available: http://www.indianrail.gov.in
6. E. Agichtein, S. Lawrence, L. Gravano, Learning search engine specific query transformations for question answering, in *Proceedings of the Tenth International World Wide Web Conference*, 2001
7. AskJeevs, website [Online]. Available: http://www.ask.com
8. WebCriteria, in http://www.webcriteria.com, website
9. AnswerBug, in http://www.answerbug.com, website
10. START, in http://start.csail.mit.edu/, website
11. PRWeb, Happy VAS customers translate directly into increased ARPU and profit, in http://www.prweb.com/releases/2003/2/prweb58656.htm, 2003
12. YellowLine, Infomedia yellow pages, in http://www.yellowpages.co.in/, 2004
13. JustDail, Talking yellow pages, in http://www.justdial.com/, 2004
14. AltaVista, in http://www.altavista.com, website
15. E. Agichtein, S. Lawrence, L. Gravano, Learning search engine specific query transformations for question answering, in *Proceedings of the Tenth International World Wide Web Conference*, Tata Infotech Limited, 2001, in http://www.tatainfotech.com, website
16. MCGM, in http://portal.mcgm.gov.in/irj/portal/anonymous/qlcomplaintreg. Accessed Feb 2008
17. I. World, in *Electronic Municipal Information Services—Best Practice Transfer and Improvement Project*. Accessed Feb 2008
18. B. Chaumon, S. Guermond, Study of conditions of use of e-services accessible to visually disabled persons, in *Proceedings DEGAS 2007*
19. T. M. C. of Greater Mumbai, in http://portal.mcgm.gov.in. Accessed Feb 2008
20. W. News, in http://wirelessfederation.com/news/category/mobile-penetration/. Accessed Feb 2007

© Springer Nature Singapore Pte Ltd. 2018

P. V. S. Rao and S. K. Kopparapu, *Friendly Interfaces Between Humans and Machines*, https://doi.org/10.1007/978-981-13-1750-7

21. D. Hong, An introductory guide to speech recognition solutions, in *Industry White Paper by Datamonitor*, 2006
22. M. Research, Microsoft speech—solutions: password reset, http://www.microsoft.com/speech/solutions/pword/default.mspx, 2007
23. Tellme, Every day info, http://www.tellme.com/products/TellmeByVoice, 2007
24. R. Pieraccini, C.-H. Lee, Factorization of language constraints in speech recognition, in *Proceedings of the 29th Annual Meeting on Association for Computational Linguistics*, Association for Computational Linguistics, Morristown, NJ, USA, 1991, pp. 299–306
25. S.L. Young, A.G. Hauptmann, W.H. Ward, E.T. Smith, P. Werner, High level knowledge sources in usable speech recognition systems. Commun. ACM **32**(2), 183–194 (1989)
26. D. Buhler, W. Minker, A. Elciyanti, Using language modelling to integrate speech recognition with a flat semantic analysis, in *6th SIGdial Workshop on Discourse and Dialogue*, Lisbon, Portugal, Sept 2005
27. V.W. Zue, J. Glass, D. Goodine, H. Leung, M. Phillips, J. Polifroni, S. Seneff, Integration of speech recognition and natural language processing in the MIT voyager system, in *Proceedings of ICASSP*, vol. 1, 1991, pp. 713–716
28. Y.-Y. Wang, A. Acero, M. Mahajan, J. Lee, Combining statistical and knowledge-based spoken language understanding in conditional models, in *Proceedings of the COLING/ACL on Main Conference Poster Sessions*, Association for Computational Linguistics, Morristown, NJ, USA, 2006, pp. 882–889
29. S.K. Kopparapu, A. Srivastava, P.V.S. Rao, Minimal parsing keyconcept based question answering system, in *HCI (3), ser. Lecture Notes in Computer Science*, vol. 4552, ed. by J.A. Jacko (Springer, 2007), pp. 104–113
30. S. Pankanti, A.K. Jain, Integrating vision modules: stereo, shading, grouping, and line labeling. IEEE Trans. Pattern Anal. Mach. Intell. **17**(9), 831–842 (1995)
31. C. Fellbaum, English verbs as a semantic net. Int. J. Lexicogr. **3**(4), 278–301 (1990)
32. S. Kopparapu, A. Srivastava, P.V.S. Rao, Minimal parsing keyconcept based question answering system, human-computer interaction, in *HCI Intelligent Multimodal Interaction Environments*, pp. 104–113, LCNS Vol. 4552, 2007, Springer, ISSN: 0302-9743 (Print) 1611-3349 (Online)
33. S.K. Kopparapu, Voice based self help system: user experience vs accuracy, in *Innovations and Advances in Computer Sciences and Engineering*, 1st edn, ed. by T. Sobh (Springer, 2010), ISBN-13: 978-9048136575, March
34. S. Kopparapu, A. Srivastava, P.V.S. Rao, A natural language interface for a railway website, in *Second National Conference on Innovations in Information and Communication Technology 2006*, PSG College of Technology, Coimbatore, 24 Apr 2006
35. S. Kopparapu, A. Srivastava, P.V.S. Rao, KisanMitra: a question answering system for rural indian farmers, in *International Conference on Emerging Applications of IT (EAIT 2006)*, Science City Kolkata, 10–11 Feb 2006
36. S. Kopparapu, SMS based natural language interface to yellow pages directory, in *Mobility Conference 2007*, Singapore
37. https://en.wikipedia.org/wiki/Question_answering

Printed in the United States
By Bookmasters